BOSTON FIRSTS

BOSTON FIRSTS

*40 Feats of Innovation and Invention That Happened
First in Boston and Helped Make America Great*

LYNDA MORGENROTH

Beacon Press, Boston

Beacon Press
Boston, Massachusetts
www.beacon.org

Beacon Press books
are published under the auspices of
the Unitarian Universalist Association of Congregations.

19 18 17 16   8 7 6 5 4 3 2

Text design by Patricia Duque Campos
Composition by Wilsted & Taylor Publishing Services
Illustrations by Joel Holland

Library of Congress Cataloging-in-Publication Data
Morgenroth, Lynda.
Boston firsts: 40 feats of innovation and invention that happened
first in Boston and helped make America great / Lynda Morgenroth.
p. cm.
ISBN 978-0-8070-7134-2 (paperback   )
1. Boston (Mass.)—History—Miscellanea.
2. Inventions—Massachusetts—Boston—History—Miscellanea.
3. United States—Civilization—Miscellanea. I. Title.

F73.36.M67   2006
974.4'61—dc22   2005031729

To Emily
Colleague and Friend

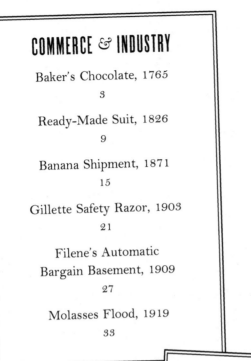

Preface

The world of Firsts you are poised to enter is surprising (who would think of sending parts of Boston ponds to Calcutta?), sobering (the murderous Molasses Flood could have been prevented), and hilarious (imagine breeding a fighting dog that wound up as a little sweetheart, googly-eyed in your lap, licking your hand). But these forty Firsts are not comprehensive. We did not begin on the very morn our civic ancestors arrived, proclaiming this City on a Hill, and then proceed to track every invention, discovery, amendment, remarkable mishap, singular accident and incident, and canny legislative maneuver, from 1630 to the present moment. We were selective. We were discerning. We became connoisseurs. We selected Firsts to convey a range of endeavor—from science and commerce, to art and architecture, to publishing, government, and nature.

Though liberal in selection, we—which is to say, I and research assistants Melissa Carlson and Elizabeth Steffey—were conservative on facts, restraining our enthusiasms. Many much-vaunted Firsts turned out to be Seconds, or Thirds, or even tawdry Fourths. Gnarly myths, hard to disentangle, had grown up around many, but with steely eyes and strong pitchforks we three researchers uncovered hidden roots. The Firsts we chose had to be in an area that was part of Boston at the time of the event's occurrence. Revere Beach, for example, endlessly cited as Boston's first public beach, is located in Revere, which only belonged to Boston from

1634 to 1739. The beach opened in 1896, and today Revere is its own city. The burning of the Ursuline Convent—a fire set by a mob of Protestant men, the subject of the book *Fire and Roses*—occurred in Charlestown, in 1834, before it was annexed by Boston in 1874. Many Boston inventions, upon closer analysis, occurred in Cambridge laboratories, inventors' minds invigorated by the easterly breeze from across the Charles.

Occasionally we zeroed in narrowly on what constituted "firstness," the Public Garden being one example. An earlier urban botanical garden existed in New York City, but it was never a civic enterprise in the way Boston's Public Garden was. The land was privately owned, and the creator acted mainly on his own, unlike occurrences in Boston, where the people of the city harangued the legislature for decades, and clergymen conducted stunts such as parading schoolchildren festooned with ringlets to promote the Public Garden. We take no pleasure in New York's loss, but even the memory of that public garden is crushed beneath Manhattan real estate. Alas, New York lacked a clergyman with a green thumb and a talent for PR.

The roster also contains a few selections that emerged in the mind and opinions of yours truly. Take Isabella Stewart Gardner, doyenne of beauty and variety. The great lady needed no further citation for contribution of Firsts in her collection (first Matisse in America, first Raphael) and for almost single-handedly bringing Italian art and architecture to the US of A. But it was as a cultural activist and avatar that I personally chose to portray her. Why? Author's call, a matter of perception and interpretation, though the facts substantiate the call.

We researchers didn't get obsessive (well, maybe a little) but tried to represent five centuries—from the Boston Common (1634) through the Puritans' ban on Christmas (1659), to the controversial first inoculation for smallpox (1721), to the founding of the first YMCA (1851) in America, to the creation of Filene's Automatic Bargain Basement (1909), to the legalization of same-sex marriage (2003). Boston Firsts contains a big crop of nineteenth-century Firsts because it was the age of invention, and a sparser harvest during the twentieth century because we thought readers would find the past juicier. When Firsts led to other Firsts, and correspondences developed between Firsts separated by centuries, they were hard to resist and guided certain selections. Who could resist the first ready-made suit, which led to the founding of Simmons College, the first college in the U.S. to combine liberal arts and professional studies for women? And how, in this city of the Common, could we ignore a new town common, near new row houses—the form Charles Bulfinch brought to Boston two hundred years ago—especially since the new row houses went up in Roxbury just ten years ago, on an old street named for Puritan governors...?

Boston being a bookish community today, as in the past (first novel, first newspaper), there were the none-too-easy issues of tone and approach. To be serious or humorous in the telling of tales; to aim for long-range impact or short-range merriment; to design the book for a party, a reading over breakfast, or a seminar? Readers will find one and all, a range of treatments, because Boston history dictates such. You will soon read of the magnificent achievement of a physician whose research and precedent-setting surgery launched the practice of organ transplantation, and of the first row houses on our shores, carried from England in the mind of an architect. You will find Puritans peering in windows at sexual goings-on, marshmallows sold with safety razors, and a bowl of creamy risotto being delivered—not unlike a pizza—from a ritzy Boston hotel to Isabella Stewart Gardner at Fenway Court.

Please join us. You'll be as amazed as we were.

Lynda Morgenroth
September, 2005

COMMERCE
& INDUSTRY

BAKER'S CHOCOLATE

❦ 1 7 6 5 ❦

There's an old idea about Boston, in particular early Boston, that its inhabitants were joyless, repressed, and dull. What a narrow view, as though pleasure were a recent invention. People—we spirited, willful, pleasure-seeking creatures—are not very different anywhere or anytime. A perusal of eighteenth- and nineteenth-century history reveals a great capacity for pleasure and its unquenchable, luscious pursuit.

Boston had chocolate—the exotic bean, the seductive drink, the first, finest, most long lasting chocolate mill in America. Baker's Chocolate—the same shiny squares we still use to make devil's food cake, chocolate pudding, brownies—set up shop in 1765 and continued to scent the Neponset River region of Dorchester until the factory was moved to Dover, Delaware, in 1965.

A society that supports the manufacture, sale, and aromatic slow melting of chocolate—mixed with sugar and rich milk to make a beverage—or (following the requisite technical developments) applied directly to the tongue in the form of a creamy bonbon, cannot be a society without pleasure and joy.

The pleasure-bestowing mill came about because of a chance meeting between two men on a road in 1764. Dr. James Baker, a Dorchester physician, perhaps knowing more than most about pleasure and its pursuits, and having capital to invest, strategized and financed the mill. He didn't know beans about making chocolate—coaxing the brittle beans into an elixir—but he met up with an artisan of chocolate, John Hannon, who had learned the trade of processing chocolate in London, where the drink was as fashionable as coffee.

Baker was Harvard-trained, confident, prosperous. Hannon was an Irish immigrant, down on his luck. Both men were canny, inventive, ambitious, and loath to leave a good idea alone. They were inhabitants of Dorchester (which became part of Boston in 1870), out and about on foot. Hannon had come to America looking for a better life. Baker, who already owned a store in Boston, was interested in employments beyond medicine.

Perhaps the day was sunny, breezy, propitious, leading them to linger and chat. Perhaps it was windy and cold, and the windblown strangers exchanged wry comments on the hellishness of the weather. They may have sensed an energy about each other— a restlessness, game spirit, zeal for enterprise.

By the following year, they had set up a manufacturing business, the Baker-Hannon factory. Dr. Baker had found an ideal site, an old sawmill on the banks of the Neponset River, which provided the water power needed to run the mills to crush the beans. Their product caught on, creating a stir in taste-conscious Boston households. It seemed a miracle that such an ambrosial, exotic drink could be made from a small, dark-brown cube—that the pleasures of taste, comfort, and stimulation could be so reliably attained.

Regrettably, in the eighteenth century, even the most domestic-seeming ventures posed dangers. At the very start of the company's two centuries as a local enterprise, chocolate maker John Hannon disappeared at sea in 1779, en route to the Caribbean to procure cocoa beans. Some said that he was trying to escape his wife. Dr. Baker wrangled with his widow for ownership of the factory. Mrs. Hannon is described as difficult, but she may only have been demanding her due. The name of the mill became the Baker Company. It passed from father to son, and in 1820 to grandson Walter Baker.

In eighteenth-century terms, it was neither coincidental nor odd that an estimable physician started a chocolate company. Chocolate was considered an aid to digestion, a health food. It was recognized as a stimulant—a quick, concentrated source of energy —and accorded aphrodisiac powers. Aztec king Montezuma, with whom explorer Cortés first shared a cup, was said to have been a sexy fellow, fueled by the consumption of fifty large cups of chocolate each day—this without benefit of sweeteners, a European innovation. (The Aztec word for chocolate, *xocolatl*, means "bitter water.")

The beans were brought by the seventeenth-century Spanish explorers from the New World to Europe. There, chocolate became a fashionable beverage, the latte of its day, dispensed in shops favored by fashionable young men—"bloods," as described by Marcia and Frederic Morton in *Chocolate: An Illustrated History*. The caffeine-soused gents chatted of affairs weighty and frivolous over their cups. Hot chocolate was also prepared by servants in wealthy households, carried in small china pots on trays into bedrooms.

Yankee traders had started bringing in beans from the West Indies before Dr. Baker's enterprise, but in the American colonies, there had been only one short-lived attempt at manufacturing a usable product. The pleasures of chocolate we savor today were just the same two centuries ago—men, women, and children were captivated by its luxurious richness, but the richness was difficult to extract. Even in an era when most foods were a pain to prepare— with the possible exception of an apple—chocolate was a royal pain. It had to be imported from equatorial lands and kept from rotting. Each bean had to be extracted by hand from a pod (in which there were a few dozen beans), dried, roasted, and then, with Herculean force, crushed with the apothecary tools of mortar and pestle.

As a physician, Dr. Baker had probably seen his share of "chocolate elbow," the stress injury suffered by the family servant assigned to crush chocolate beans by hand. (In Europe, where chocolate drinking was more advanced, itinerant artisans went from house to house with grinding wheels, sparing the kitchen servants from sprains.) Using a mortar and pestle, even with an impressive show of force, the best one could hope for was an irregular mash. The resulting chunks, chips, and smidgens were of inconsistent flavor and texture, though the aroma of chocolate did mercifully come through. The pulverized beans were then smooshed into a paste with sugar, mixed into milk, and heated.

By contrast, Hannon's Chocolate was convenient, clean, handy, and neat—a brick-shaped bar, and later a cube, that cooks could scrape down to prepare hot chocolate. The bars were expensive because of the tariff levied by the British, the same hated tariff that affected molasses and sugar. But preparation became practicable and pleasant. An enlivening aroma scented the drafty kitchen and filtered out to the dining room and upstairs to bedrooms and breakfast rooms, as chocolate from the tropics simmered on Boston stoves.

The enterprise expanded along the Neponset River and came to include a gristmill and cloth mill created by Dr. Baker's son, Edmund. The younger Baker was prescient and enterprising; in the midst of the War of 1812, when the British blockade of Boston prevented beans from entering the city for two years, he seized the opportunity to build a bigger, better mill. Following the war's end, business soared.

The company became a major operation, involving international trade and transport, the employment of a variety of skilled and un-skilled labor, and the use of sophisticated marketing. During the nineteenth century, Walter Baker—Edmund's son, grandson of Dr. Baker—pioneered in the use of advertising, including placement of the comely "Chocolate Girl" by Swiss artist Jean-Etienne Liotard on packages, tins, and ads.

Baker Chocolate pioneered in business, introducing a money back guarantee—"If the Chocolate does not prove good, the Money will be returned"—which became an American tradition. They hired women employees, starting in the 1830s and 1840s, advancing the practice of employing "factory girls" in New England mills. But the company was no model of worker treatment. Conditions were squalid—sweltering heat in summer, drafts and bitter cold in winter—according to Peter F. Stevens, in a recent article in the *Dorchester Reporter.*

In 1927, General Foods bought the family enterprise, and in 1965, the Dorchester factory was closed. Production was moved to Delaware. The handsome red-brick buildings, part of a chocolate village on the Neponset—which looked like an illustration in a Victorian children's book, with mansard roofs and elaborate trim— were vacant for many years, a standing offense to the enterprise of the founding family. They have since become exclusive condominium apartments, a form of profitable investment Dr. Baker would have found greatly to his taste.

AUTHOR'S NOTE

Urban sleuths, follow Dorchester Avenue from the Boston neighborhoods of South Boston and Dorchester, and you'll find the grand, Second Empire–style buildings at the Dorchester/Milton line. The factory neighborhood has become chi-chi, with antique stores and restaurants. But vestiges of an older, bookish Boston linger, including the Lower Mills branch of the Boston Public Library, offering books to the millworker since 1876.

READY-MADE SUIT

In the annals of Boston Firsts, there are several notable twofers and even threefers, firsts that beget firsts that beget firsts. Typically, these are technological. Over time, a first lighthouse would lead to the invention or adaptation of innovative lenses and lamps. The first ice cut from New England ponds—swathed in straw, loaded onto ships, delivered to tropic climes—would lead to the development of icebox and refrigerator firsts. Ether used in anesthesia would require clever devices to turn liquid into gas.

But occasionally, a Boston First leads to a totally different kind of first, as though one planted a tomato seed and a pineapple tree grew! Regarded more closely, what has actually happened is that one man or woman's inventiveness, insight, and derring-do are at work in multiple milieux.

Such is the case of John Simmons (1796–1870). His diverse accomplishments now seem logical, even ordained. But during his lifetime, it would have seemed unlikely that a dour merchant of Pilgrim origins would become a feminist. Simmons exemplifies the potential of a man attuned to his time, or sufficiently ahead of it to self-propel, prosper, and contribute to society, including after death.

He was a true son of New England, a descendent of Moses Symonson, who arrived in Plymouth, Massachusetts, on the aptly named ship *Fortune* in 1621. Simmons (the name's spelling was changed by Moses's son, John) grew up in a house that still stands in the Sakonnet district of Little Compton, Rhode Island—a small, shingled eighteenth-century cottage moved from West Main Road to Simmons Road during the twentieth century. Simmons went to Boston when he was eighteen years old to apprentice in the tailor shop of his older brother, Cornelius. The brothers both sewed for their supper, but John took particular notice of the changes on the street. For the retailer and manufacturer he would become, the street they worked on was an ideal test market. Ann Street (today North Street, running just beside the Callahan Tunnel) was the hub of the clothing business in Boston.

Alert to his environment and a young man himself, Simmons noticed the increasing number of clerks and apprentices bustling about on Ann Street. At age twenty-two, he opened his own shop on Ann Street and got the bright idea to make ready-made suits in standard sizes, the ancestor of all our off-the-rack rags. Not only were these garments far more affordable than custom-made suits —previously, the only game in town—but the sartorially deprived might order from afar. Once, twice, and more. This was a boon comparable to Internet shopping. Simmons moved a few times more on this lively, crowded thoroughfare. In 1825, the Boston Directory listed thirty-five stores as purveyors of ready-made clothing, writes Harry Corbin in *The Men's Clothing Industry;* twenty-one of these shops were on Ann Street.

Simmons sent well-dressed salesmen, armed with swatches and samples, into the American South, the West, the boonies of New England, and all along the Eastern Seaboard. In this era of enterprise, traveling salesmen seemed almost to wait in line for new railroad routes and extensions, hopping on board at first opportunity.

During the early years of the Simmons business, suit construction was farmed out to the homes of poor women all over Boston. Work was done by hand. This was a pre-sweatshop era, as the sewing machine had not yet been invented. Simmons would go from tenement to tenement, room to room, delivering the cloth that had been sized and cut by his tailors for the homeworkers to assemble. It is not known how Simmons responded "in the moment," as we would say today, but his later actions suggest that he was moved and disturbed by the plight of the women he encountered. Their living conditions and prospects were often miserable. Even by the mid-nineteenth century, it was not considered respectable for women to work outside their homes. And yet 15 percent of the labor force were women—mainly shopgirls, domestics, and sewing women, according to Claire Golding, writing of John Simmons in the 1974 *Simmons Review.* Many seamstresses were desperate—farm girls turned city girls totally on their own, or widows with young children, or older women struggling to support impoverished elderly parents. These women toiling at their poorly lit kitchen tables contributed to John Simmons's fortune.

In 1826, he left Ann Street forever and expanded to occupy the second floor of the Quincy Market Building—centerpiece of today's Faneuil Hall Marketplace—to mass-produce ready-made suits for men. Orders soared. Prosperity followed. By 1875, almost thirty thousand people in Massachusetts were employed in the manufacture of ready-made clothing, about fifteen thousand in Boston alone, according to Justin Winsor, author of *Memorial History of Boston.*

Simmons built fancy brownstone houses for himself and his family, one facing Boston Common and others on Arlington Street. In 1851, for the first time, he had a building especially designed and constructed for his enterprise. The grand granite "Simmons Block," a landmark at the corner of Water and Congress Streets, included a store, The Clothing House, on the ground floor, with offices above and a thriving clothing factory on the top floor.

Buying real estate felt good—solid and profitable—to Simmons; building on it, adding value, felt even better. During the next sixteen years, he bought choice downtown property, erected handsome, high-quality buildings, and carefully managed them, adding still more value. By 1867, he was one of Boston's wealthiest men.

During his lifetime, Simmons did nothing more flamboyant or

uncustomary with his money than invest in real estate—more and more of it, especially in the vicinity of today's Post Office Square. When he died in 1870, his will revealed that, with his lawyer, he had carefully and assiduously designed a plan to found a college for women, a place that could set them free. He directed his trustees to keep his principal invested in real estate and, when $500,000 had accrued, to establish a place of higher learning that would lead women to economic self-reliance.

Reporter David Patten wrote about Simmons in *The Providence Journal* some forty years ago, describing the language of Simmons's intention. His bequest for a "Female College" was to convert seamstresses to educated women; to teach "medicine, music, drawing, designing, telegraphy [operating telegraph machines] and other branches of art, science and industry best calculated to enable the scholars to acquire an independent livelihood." When Patten wrote his article, Simmons alumni numbered about thirteen thousand; a *Providence Journal* copyeditor titled Patten's piece, "S'cunnet Native had 13,000 Daughters."

And so the essential story is simple, symmetrical, yet surprising. A savvy small-town fellow comes to Boston, becomes a tailor, sizes up the situation, sees the need for affordable and accessible gentleman's suits, and supplies them. In the process, he becomes a wealthy man and turns philanthropic. He remembers the desperation of the women who pieced together the suits he sold, and is determined to do something to advance them. But instead of starting a fund for the impoverished, or donating to an existing agency, he determines to create a college for women that will make them self-sufficient.

Perhaps it was not only the plight of the women that moved John Simmons, who had two daughters of his own. He suffered greatly in his own life, despite his wealth. He started out as a poor boy, and married Ann Small of Providence, Rhode Island, when very young. They both struggled and had six children. Accounts differ, but most likely all four of John Simmons's sons died, as well as his single, much beloved grandson, ending the family line that had started with Moses Symonson's arrival in 1621.

Misfortune followed him after death, but even then, his careful management triumphed. In 1872, two years after Simmons died, the Great Boston Fire destroyed everything between Washington Street—near today's Downtown Crossing—and the waterfront. Simmons had put all his money into real estate, and with it his hopes of starting a college for women. The fire destroyed every one of his properties. But Simmons had chosen his trustees wisely. They borrowed money and built upon the land where the buildings had once stood, paying off the mortgages for nonexistent buildings (he had carried no fire insurance). By 1899 they'd saved enough to make good on his wishes. Simmons College opened in 1902, the first college in the U.S. to combine liberal and professional studies for women.

For over a century, this college on the Fenway near Boston Latin School and Mrs. Jack Gardner's Italian palace (see page 123) has been a pioneering force in the lives of tens of thousands of American women. John Simmons also helped to establish a New England tradition for education-directed philanthropy among clothing manufacturers.

Perhaps not coincidentally, in modern times, Simmons College became legendary for its women-focused MBA program, along with its successful programs in social work, nursing, library science, and publishing.

To update the old *Providence Journal* headline, it would now need to be said that John Simmons had not two, nor thirteen thousand, but thirty thousand daughters. Many of them wear ready-made suits.

BANANA SHIPMENT

Nowadays, all kinds of high-flown concepts and pricey consultations are associated with developing new products. Centuries ago, it was as simple as two men meeting on a road and chatting—as in the case of the origins of Baker's Chocolate—or young men daydreaming about moving ice from Boston ponds to parched ports in India. Then as now, specialties, even empires, were born of inspiration, insight, ingenuity, and finding a likely partner. In New England many great ideas originated in filling gaps, figuratively and literally—including the empty holds of ships. A seaman would be hired to carry cargo in one direction, but not have anything to return with.

Which brings us to Captain Lorenzo Dow Baker and the banana, our cheerful, everyday, lunchbox fruit—once hidden in the mists of tropical forests, and unknown to North Americans.

On a fateful day in 1870, Lorenzo Dow Baker (1840–1908), a sea captain from Wellfleet, Massachusetts, who had just delivered mining equipment to Venezuela and picked up a shipment of bamboo in Jamaica, chanced upon a banana in a Jamaican market. He bought one, ate it, loved it, marveled at its golden yellow peel and delicate pale yellow interior. Captain Baker was always scouting about for novelties and new products, especially when he had emptied the hull of his schooner, *Telegraph*, and would otherwise have to return to homeport without cargo—a waste of potentially profitable space. He brought a sample of nicely ripe bananas to his next port of call, New Jersey (or New York, by some accounts). The fruit rotted en route. But Baker, an experienced seaman—well able to remember and to forecast—had not forgotten the luscious taste of the banana, nor the period of time it took ripened fruit to rot. Like the backs of camels, and for similar reasons, the temperature and motion of ships can be beneficial to foodstuffs. Captain Baker thought there might be a future for the slow, steady ripening of bananas on board the *Telegraph*.

He also noted the shape and growth habits of the fruit. A "hand" of bananas—the tidy grouped array plucked from a tree—seemed ideal for shipping, each curved fruit nestled in an adjacent cushioning fruit, and individually wrapped in a thick, insulating, waxy peel that would protect it during a long voyage. At the start of the voyage, the peel was the thickest. Near the end of the voyage, the peel thinned, releasing an aroma that fruit lovers found intoxicating.

The very next year, 1871, the captain/fruit-explorer loaded unripened bananas into the hold of the *Telegraph* and made his way to Boston. He arrived at Long Wharf—checking in at the imposing, granite Custom House Block built by Isaiah Rogers (see page 137) —with about four hundred fragrant, ripe bunches of bananas.

The captain's exotic, handlebar-shaped fruit was a sensation. Boston purveyors were eager to buy it, even before they tasted it, because its appearance was so novel; its display in a shop's window at once seduced customers into a grocer's lair. Fruit companies were anxious to carry it because of the panache associated with tropical fruit; only the best companies could boast of exotics from afar. This was an era long before large-scale growing in California and Florida—there were no California and Florida as we know them—and before large-scale refrigeration.

As this tale of fruitful commerce evolved, Captain Lorenzo Dow Baker, who became known as the Banana King, would require the innovative contributions of Frederic Tudor, the Ice King (see page 167; but we are getting ahead of the story), who made refrigerated cargo possible. For the next decade, the captain brought in bananas from Jamaica, acquiring new schooners for his proud fleet and adding coconuts to his tropical larder.

Captain Baker's Boston banana introduction was the first commercial shipment of bananas into the U.S. and would lead, both wonderfully and terribly, to the founding of the United Fruit Company in 1899. In its ultimate evolution, or devolution, the company would become a virtual colonial power in its exploitation of Latin American laborers. It was called *El Pulpo*, "the octopus," by native workers.

Enter Andrew Preston (1846–1924), one of the eager buyers of Captain Lorenzo Dow Baker's first Boston bananas. Preston, a young man out of Beverly, worked at a Boston produce firm. From the very beginning, and whenever Captain Baker arrived with a shipment of bananas, Preston would buy them for his clients. A rapport developed between captain and purveyor, fruit explorer and fruit marketer. During the 1880s, the Boston Fruit Company was established by Captain Baker and Jesse Freeman, a Wellfleet merchant, with Andrew Preston as junior partner. As the consumer demand for bananas escalated—one is tempted to say that America went bananas—the company was incorporated and, in 1888, Preston made general manager. Bananas became the signature product of Boston Fruit. During the 1890s, Preston— who envisioned an empire—bought up land and developed sprawling banana plantations in the Caribbean and Central America, transported the fruit in state-of-the-art refrigerated steamers, and created a refrigerated distribution system to move and sell the bananas across the U.S.

Many banana eaters, then as now, imagine bananas growing wild in the tropics, with monkeys climbing about and colorful birds alighting on the thick, ragged leaves. In fact, all bananas produced for human consumption are grown on plantations; the variety we eat is seedless and must be hand-planted and tended. Another myth is that native people wait until bananas ripen on trees, then pick them. Like the exporters, natives also pick the fruit when it's green and allow it to ripen off the tree.

The banana itself becomes a character in this saga of exotic food supply, so suitable is the fruit for shipping, handling, and commercial enterprise. (And who could be uninfluenced by the comely Miss Chiquita Banana, United Fruit's 1944 creation?) The banana takes weeks to ripen and—as Captain Baker quickly and profitably discovered—the long ocean voyage, gentle rocking of the vessel, and storage with masses of other bananas (promoting the pooling of gas that ripens fruit) create perfect incubators. Ripening can be impeded with refrigeration. Picked green, the fruit is sturdy and needs no wrapping. By the time it has become softer, more fragile, the worst jostling and danger to its ripening pulp are over.

The banana not only saved the day for Captain Baker, augmenting his cargo on those early voyages, and made the day for produce man Andrew Preston—turning him and Captain Baker into wealthy men—but rescued and ultimately favored another entrepreneur.

As Baker, Freeman, and Preston were building Boston Fruit, Minor Cooper Keith (1848–1929), scion of a New York railroad-building family, was virtually invading the interior of Costa Rica to build a railroad from San José in the interior to Limón, the country's Caribbean port. For this swashbuckling enterprise he hired thousands of workers. Hundreds died cruelly in the jungle. On contract to the Costa Rican government, Keith faced so many vicissitudes—from disease and loss of life to nonpayment by the government—that he started growing banana trees along the path of the railroad to feed his workers. Though not intended to be beautiful, the path from Limón featured a grand allée of banana trees! In 1873, two years after Lorenzo Baker had brought the first shipment of bananas to Boston, the Costa Rican government defaulted on payments, and the new railroad still lacked passengers. Desperate or prescient, or some combination thereof, Keith decided to export the bananas from the trees along the underused railroad.

This was a runaway success—consumers lusted for bananas—and soon Keith was making much more money from selling bananas than building railroads.

During yet another crisis in the 1890s, Keith lost all his money and, seeing the writing on the wall, traveled to Boston to meet with his longtime rival, Boston Fruit. In 1899, Andrew Preston—the boy out of Beverly turned fruit magnate by Captain Baker's bananas—and Minor Keith merged companies, forming United Fruit. It was not a venture favored by Baker, who was at heart a seaman and trader, neither profiteer nor rampaging capitalist. As president of Boston Fruit, Baker was involved in the merger that became United Fruit, but over the years he withdrew from the enterprise as it turned ugly and monopolistic. History professor Randolph Bartlett of Cape Cod Community College has done primary research into Lorenzo Dow Baker's life—gaining access to Baker's papers, held by his Cape Cod grandson—and written and lectured on the sea captain's ethical entrepreneurism.

The introduction and marketing of bananas to Boston showcases the sagacity of seamen and fruitmen. The rapid response of the public reveals a capacity for pleasurable consumption and a willingness to try new foods. Contrary to the reputation of Boston diners as culinary conservatives, the banana caught on fast. Consider that in the early 1870s, cooks had to be instructed by domestic encyclopedias how to open and eat the foreign fruit. By the late nineteenth century, the banana had entrenched itself in the American kitchen, earning a secure spot in Fanny Farmer's *The Original Boston Cooking-School Cookbook*.

Miss Farmer's book provides two recipes for banana fritters. One dainty version macerates halved bananas in sherry and lemon juice before breading and deep-frying; another is served with lemon sauce. Her Banana Salad features banana slices and cubes, marinated with French dressing, which are then put back into the banana skins and served, preferably grouped around a lettuce mound.

By the turn of the twentieth century, the no-longer-foreign fruit had earned a place in the iconography of daily, all-American menace. The first *Boy Scout Handbook*, published in 1911, provides a list of reliably good deeds, with particular attention to scooping up banana peels from the sidewalk.

It all began with Captain Lorenzo Dow Baker's sighting of, attraction to, and appreciative tasting of the bright yellow fruit. And his canny recognition that the hold of a ship might be a gigantic, gentle ripening chamber—a very big bowl for fruit.

GILLETTE SAFETY RAZOR

ᔥ 1 9 0 3 ᔦ

S tropping. Most of us don't even know the meaning, much less the motion. Once, not so long ago, men shaved with razors that looked like mini-cleavers—raw-edged blades that had to be continually sharpened. Stropped, to be precise. You lathered up, took the blade, and batted it back and forth on leather to sharpen the edge. (If you were negligent or inept in stropping, you would have to resort to honing, serious honing, which was expensive and inconvenient, as it required a professional sharpener. Then, of course, you could not shave while the blade was being honed—with a stone—though at least you had a brief respite from stropping.) The process of shaving was, essentially, running a knife at an angle to one's face and scraping. Men were always cutting their faces, especially around curves. Chins and necks were a mess. No wonder barbers doubled as surgeons! Surely there was something safer. But there wasn't. Not till King Camp Gillette. He was a man on a mission to create a new object and to get rich. Meanwhile he was selling bottle caps.

Gillette was born in Wisconsin and grew up in Chicago. Invention and high energy ran in the family. His father, George Wolcott Gillette, a hardware wholesaler, and his two brothers each had several patents. His mother, Fanny Lemira Camp, was a talented and eclectic cook who collaborated with a former White House steward, Hugo Zieman, and published *The White House Cookbook* (1887) by Mrs. F. L. Gillette. A bestseller, it featured recipes, household tips, and counsel.

At age seventeen, the personable Gillette became a traveling salesman and went to work for William Painter's Baltimore Seal Company. Painter, the company president, had invented the crimped-edge "crown cap" that most of us grew up with, the trusty metal bottle cap on root beer and cola drinks in the pre-can, pre–juice box era. Gillette admired the cleverness and confidence of his boss, and determined to come up with his own essential everyday object. He traveled about musing on categories of potential products—food, hardware, transportation, communication, grooming. He developed a mnemonic to stimulate his inventiveness, repeatedly reciting the alphabet to try to generate ideas.

Gillette's brainstorm was not an invention akin to that of Frederic Tudor, who realized that ice in ponds was a cold crop waiting to be harvested and shipped. Nor was it an invention of something new—transmitting voices across wires or making prefab suits for the masses. Gillette's invention was based on wanting to invent something, anything—cheap, necessary, and ordinary—that vast numbers of people would buy, discard, and buy more of. Forever.

In 1895, when King Camp Gillette was forty years old, still traveling, still selling—Boston being one of his territories—he was holding a dull razor in his hand one morning, and realized it was beyond the point of stropping. How irritating to a traveling salesman, who took pride and pleasure in his well-groomed appearance, which helped to open doors. Gillette had the essential idea of a thin, flexible blade in a clamping holder, exposing two sharp edges of steel, and a handle screwed into position on the clamping plate.

From a standpoint of generating capital, what could be better than an invention that removes unwanted hair, which keeps on growing as long the customer is alive and spending? "I have got it; our fortune is made," Gillette wrote to his wife in Ohio. He rushed out to Wilkinson's, a hardware store on Washington Street in Boston—shaved? unshaved? ill shaved? no one knows—and bought components to make a crude model.

The double-edged razor in a holder was far from an immediate success. Eight years elapsed before it was brought to market.

The story of this particular First, the razor with disposable blade, is another tale of a propitious meeting and a prosperous marriage between a charismatic idea-man and a gifted, diligent machinist. After several years had elapsed, Gillette met William E. Nickerson, an M.I.T. graduate, at a dinner party in Brookline. Nickerson created an effective safety razor prototype. A charter was drawn up. A lease was signed (world headquarters was over a fish market on Atlantic Avenue). Production began.

Some brief beard background: We have the notion of men in beards for most of history, including during the nineteenth and early twentieth centuries. Not so, according to social history conveyed in *King Camp Gillette: The Man and His Wonderful Shaving Device*, by Russell B. Adams Jr. By the 1880s, most men in the U.S. were beardless and would have welcomed being clean-shaven every day. But as the process was laborious and dangerous, they shaved just a few times a week, and looked it.

In 1903, fifty-one Gillette Safety Razors and 168 blades found their way to the marketplace. The following year, 90,844 razors and 123,648 blades—disposable, replaceable, infinitely renewable —were scooped up. In just five years (1908), blade sales soared to 13,000,000 a year; by 1917, 120,000,000.

It did not hurt the blade biz one bit that King Camp Gillette looked the part of a can-do businessman. Tall, strapping, with dark wavy hair, a commanding mustache, and a persuasive way of speaking—he was, after all, a salesman—his image was so convincing, marketers decided he should be the company icon, immortalized on blade wrappers. It is estimated that Gillette's precisely self-shaved visage was printed over 100 billion times. He was often recognized, even once by a group of Egyptians who had gathered around him as he was mounting a camel, en route to the

Pyramids. The clean-shaven Egyptian men began shouting, pointing, scraping their faces with their fingers, laughing in delight.

As the company grew, Boston—a port city and manufacturing center—proved to be an ideal headquarters. Everything Gillette needed was here: a pool of skilled and unskilled workers, advanced transportation and communication, systems for delivering raw materials, and the evolving fields of marketing, advertising, and public relations, which Gillette delved into and mastered, scoring PR coups through the twentieth century.

The company rode the patriotism bandwagon: "George Washington Gave an Era of Liberty to the Colonies. The Gillette gives an Era of Personal Liberty to all Men," read a 1906 ad. They staged an advertising mea culpa during the 1930s, openly apologizing for an underperforming model, the Kroman. They were clever, cunning, and ruthless in their giveaways of razors—in gas stations, in supermarkets, and even in the pockets of overalls. The free razors required blades the consumer had to buy. Gillette marketers were unintentionally hilarious in their decisions as to what products might be linked with freebies. Razor giveaways were even paired with marshmallows, according to remarks made by former Gillette president Joseph P. Spang Jr. at a fiftieth anniversary dinner of the business-oriented Newcomen Society. (Why marshmallows? Shaving sessions around campfires? Distribution to Eagle Scouts with fledgling beards? The rationale lies buried in marketing archives.)

Wartime is a fine time for makers of uniforms and munitions, and during World War I, the Gillette Safety Razor. When soldiers weren't being blasted in their trenches or poisoned by mustard gas, they were keeping their faces smooth. The U.S. government bought over four million razors. Soldiers got used to the idea of shaving themselves, as opposed to being shaved by a barber. When they came home, they kept their army-issued razors, and—in some combination of thrift, maintenance of routine, and superstition— they continued to buy Gillette blades. Forever.

In World War II, Gillette again did very well, not only in sup-

plying the military, but in cultivating another consumer habit. Soldiers—by then accustomed to shaving their own beards—got used
to another new idea: daily shaving. They took the idea, their army-
issued razors, and the habit home.

The worldwide company, which over the twentieth century extended into sundry additional consumer care products, became a
Boston institution. It employed thousands of workers, broadcast
the World Series (a First), and became indelibly associated with its
two Boston locations. Gillette covered the waterfront in South Boston, its sprawling, brick First Street production facility dominating
the longtime no man's land (now changing) in the Fort Point
Channel area. Corporate headquarters have long occupied the posh
upper stories of the Prudential Center, leasing about a third of the
1.2-million-square-foot tower, a signature image on the Boston
skyline.

In October 2005, a century-old tradition of Boston-based, in-
ternationally known safety razors changed. Procter & Gamble, the
Ohio-based consumer giant, manufacturer of all things disposable
and replaceable, bought Gillette. Most likely, the plant in South
Boston will remain, but the future of Gillette headquarters at the
Prudential is uncertain. What would King Camp Gillette make of
this? Before the man was a manufacturer and a magnate, he was a
political theorist and reformer, a kind of free-form socialist who
published his ideas in *The Human Drift* (1894). He did not see his
commercial enterprise in conflict with his personal philosophy,
pointing out that, realistically, he was a man living in a capitalist
society, and so, why not go with the flow?

Perhaps he would be disturbed by the Procter & Gamble take-
over, a consolidation that will make money for executives, managers, and shareholders, but cost hundreds of workers their jobs.
On the other hand, his eccentric philosophy encompassed the idea
that industrial monopolies were precursors of a publicly owned
worldwide corporation, which would own and run all means of
production. Procter & Gamble's swallowing of Gillette will create

—guess what?—the largest consumer products company in the world, a great leap toward a single worldwide corporation, though not one publicly owned.

Most likely, the King, who died in 1932, would have found it just dandy that Gillette's shaving products, Duracell batteries, and Oral-B toothbrushes will share corporate and shelf space with disposable diapers (Pampers), laundry detergent (Tide), and fluoride toothpaste (Crest)—all everyday, affordable, disposable, or consumable products that we have been taught we cannot live without and will continue to buy, use, discard, and replace. Forever!

FILENE'S AUTOMATIC BARGAIN BASEMENT

☜ 1 9 0 9 ☞

To us, Filene's Basement is a bargain. To Edward Albert Filene, it was science, philosophy, philanthropy—you might even say conservation—since this was a man who hated waste. He was born in Salem, Massachusetts, the first son of William Filene, a Polish immigrant who started a little shop in Salem, and later a little shop in Boston, and later a much bigger shop—William Filene, Sons and Company—with his American-born sons, Edward and Lincoln (Abraham Lincoln Filene).

Edward loved retailing and the study of human behavior. He got a huge kick out of knowing what people would do, how they ticked. He played to their preferences and their desire for betterment, meanwhile getting rich and devoting himself to progressive causes. Growing up in Salem, Lynn, and Boston, an immigrant's son amid Yankees, Edward understood the cult of value—a willingness to spend, but prudently, cannily, cleverly—and the quirky self-reliance of New Englanders.

He invented the Automatic Markdown, the Rosetta stone of retailing in Filene's Basement, which is officially known as Filene's Automatic Bargain Basement.

To this very day, when you see a customer looking like a human upholstery-sample book or a one-woman caravan—garments slung over her shoulders, straps of sandals, bras, and handbags looped on her wrists, handfuls of accessories with flopping tags clutched in her fists—and gazing upwards, squinting, what you are witnessing is the timeless spectacle of a shopper in search of the Automatic Markdown signs. Placed strategically throughout the poorly lighted, multilevel basement with its signature overflowing tables and narrow aisles—a kind of subterranean pushcart ambience—these signs explain The Policy. "An auction in reverse," Edward A. Filene called it:

Starting with the dated tag on every piece of merchandise, signifying when the item was put out for sale, garments and everything else are reduced 25 percent after twelve days, 50 percent after the next six selling days, and 75 percent after six additional days, a total of 75 percent—this on items generally priced at least half their original selling price. After another six days, a total of thirty selling days in all, the article is given to charity. Over 90 percent of the merchandise is out the door after twelve days, and just one-tenth of 1 percent winds up donated to charity, according to Leon Harris's great read, *Merchant Princes: An Intimate History of Jewish Families Who Built Great Department Stores.*

The fancy-schmancy "upstairs" Filene's department store and its bargain basement, one of the first in the U.S., were already in business when the Automatic Markdown was born in 1909. The Automatic Bargain Basement, which other merchants snickered at, predicting failure, didn't turn a profit for ten years. But the Filene brothers stayed with it. It saved their skins—and clothed the hides of tens of thousands of customers—all through the Depression, offsetting the losses of "upstairs" Filene's and establishing loyal customers who would find their way back during economic recovery. In *The Filenes*, author George E. Berkley notes that when President Roosevelt closed the banks in 1933, Filene's was able to meet the store's entire payroll from the cash sales of the Basement.

Such a concept! Such a deal! Only in Boston could the merchandising concept of the Automatic Markdown have been born. It was frugal, it was fair, it was equitable and fun. Brahmins and Bohemians, the wealthy and the working class shopped its depths. (The original Basement, across the street from today's Filene's—the elegant structure built in 1912, the last work of architect Daniel Burnham—was called The Tunnel Bargain Basement because it connected to the new subway station; see page 85.) Filene's Automatic Bargain Basement, diabolical and wonderful, enabled customers to buy things while conserving things; one acquired but also saved. It provided the fun of gambling without the guilt—gambling in a virtuous form. It encouraged independence and taste formation, since there were no salespeople to advise, much less harangue. And you could hardly grouse, or feel suspicious, that you were paying for fancy overhead. It was also habit forming. The markdown countdown created a sense of scarcity—not to mention anxiety—that repeatedly brought back customers who wanted to see what was new. The Automatic Markdown is "sticky," says Thomas Hine, author of *I Want That! How We All Became Shoppers*, using Internet parlance to describe the marketing atmosphere created by Filene's century-old idea.

The system was as much for Filene's buyers as its customers. It was perennially difficult to get buyers to admit they'd made a mistake and to bite the bullet and mark down merchandise. Placing the goods in the Basement took the markdown decision out of their hands.

In another era, Edward Filene (1860–1937) might have become a social psychologist (he was accepted to Harvard, but couldn't go), a profession he practiced along with several others, ranging from financier to political tactician to diplomat to social theorist, albeit without a license. He loved retailing, the whole province of acquiring and moving goods—everything from locating quality products and making them available to understanding what it was that made people buy, or refuse to. He had a lifelong bee in the bonnet about distribution. The Automatic Markdown evolved from his annoyance, indignation, and distaste for the wastefulness of conventional distribution. He sought to round up high-quality merchandise begging for buyers and deliver it.

His affection for customers extended to the most cranky and frugal, whom he particularly admired and lauded. "It is economic treason to shop carelessly," he wrote, encouraging consumers to shop in a *professional* manner, to become educated consumers and to buy judiciously.

Much unruly behavior that would not float in Boston's sedate downtown was condoned, even applauded, in the Automatic Bargain Basement. Women routinely took their clothes off in the aisles —as they still do, considering the entire space a personal dressing room—slipping garments over their heads, jostling for position in front of narrow mirrors on posts. Today, the cognoscenti shop the Basement in a leotard, the better to peel down in sleek propriety and to evaluate a garment's fit.

Edward Filene's strong social conscience led him further from both the department store and the Basement. While his brother, Lincoln, minded the store, Edward involved himself in the great causes of the day. The family was attentive to its employees as well, instituting many labor reforms, ranging from the credit union to a company union to retirement-fund stock purchases. Several managers made their mark in the Basement, and went on to attain executive status, power, and wealth.

Basement buyers were specially trained by Filene's and became virtuosos of acquisition and marketing. Regular buying arrangements (including agreements to leave the labels in) were made with upscale stores such as Saks Fifth Avenue, where "deep discounting" was heretofore thought to be damaging to the stores' prestige. In times of war, including World War II in Paris, just before the city fell, buyers rescued precious goods not merely for profit, but as a point of honor, to conserve items of value. When the British ocean liner *Queen Mary* was converted into a troop ship, Filene's buyers bought the contents of the entire ship—which contained many fine shops—and sold everything in Boston in under three hours. In 1946, Neiman Marcus fire-sale items were transported to Boston in fire trucks.

Still today, when a customer descends into the Basement, something of the elevating philosophy of Edward A. and Lincoln Filene prevails. A silk scarf rescued from oblivion floats out upon a buyer's shoulders to grace a special day; an eighty-dollar men's tie miraculously available for eight dollars ignites a new romance; a luscious, nubby, five-hundred-dollar Scottish tweed sport coat (thirty-five dollars) purchased with a shopper's last few bucks, goes forth and impresses an interviewer, wows a boss, snags a job. The Filene brothers were not averse to magic. They just liked to understand the trick.

AUTHOR'S NOTE

While Filene's Basements now exist as separate entities in suburban malls, there is only one Automatic Bargain Basement—two floors of it, a basement and sub-basement, below the original, eight-story Daniel Burnham building. You may find good buys at the other Basements, but you will not find history, mystery, nooks and crannies, ancient and venerable salespeople (especially in the men's department, tape measures slung around their necks—not that they need a tape to "know your neck"), along with the patina of accumulated human hopes and aspirations, threads of flannel and tulle, dirt and paydirt, the subway still shaking up the place, pretzels and chestnuts roasting outside, couples kissing over engagement rings, or for no reason, or because the almost-century-old place is public and private, sacred and dirty, hilarious, endearing, and sublime.

MOLASSES FLOOD

❦ 1 9 1 9 ❧

January 1919. Bostonians were going about their lives, trying to regain a sense of normalcy. But a feeling of menace hovered. People felt uneasy, much as we do in modern times when disasters —war, domestic turbulence, the scourge of infectious disease— coalesce.

World War I had just ended, the Armistice signed in November 1918. Reminders of its grisly overseas horrors were still everywhere, including the amputees and vacant-eyed vets—gassed and blind—who hobbled through Boston streets. America had only been in the war for a year and a half (April 1917 to November 1918), but forty thousand Boston boys had served. The horrific epidemic of influenza had swept the U.S. the very same autumn the Armistice was signed, killing half a million Americans—including an estimated eighteen thousand servicemen—in just two months. Sickness and death were rampant in Boston's North End, a beleaguered immigrant neighborhood packed with Italian families. Contagious disease took virulent hold in the congestion of the tenements—cold, badly ventilated buildings with poor sanitary conditions and undernourished tenants.

The North Enders, scorned and persecuted, kept mainly to themselves. The taunts against Italian immigrants—predominantly poor, uneducated Sicilians—were worse in this era of attacks by Italian anarchists inflamed by opposition to America's entry into World War I. The arrest and trial of anarchists Sacco and Vanzetti

—and the long preceding manhunt—brought local anti-Italian prejudice to an ugly pitch.

But the domestic terror that struck on January 15, 1919, was local, ordinary, from a substance innocent as butter. Around noon, a fifty-foot-tall tank of molasses—sited on Commercial Street between Boston Harbor and the tenements—burst open, emitting a crushing, suffocating fifteen-foot wave of lethal brown syrup, 2.3 million gallons of it in a pressurized rush. The bursting tank let out such a roar and crash that some terrified residents thought it was a revival of the war, the Germans bombing Boston. The viscous, inescapable tide initially traveled at speeds of thirty-five miles per hour, crushing everything in its path, even twisting the El. It pulled down houses and buildings, drowned men, women, and children, horses, dogs, and cats. They were submerged and suffocated, or swept into the frigid harbor waters and drowned, or killed by some hellish combination of fatal forces.

The scores of injured were affected in cruel and ghoulish fashion, thrown into curbs where their bones were broken and heads split, or felled by chunks of shrapnel-like steel from the exploded tank. Stonecutter John Barry, in excruciating pain, was injected with morphine in his spine three times as rescuers tried to dig him out of the collapsed brick firehouse that had stood near Boston harbor. Medical examiner Dr. George Burgess Magrath, clad in rubber hip-boots, plodded through the wreckage, identifying the crushed and lacerated bodies. Eight years later, on August 25, 1927, the same Dr. Burgess would perform the legally required autopsies on Sacco and Vanzetti, who had been electrocuted the previous day following a six-year appeal.

In the dark wake of the flood, twenty-one people were killed—longshoremen, teamsters, pavers, a fireman, a housewife, and a boy and girl, each aged ten; 150 were injured, some crippled for life. Never before had an ordinary ingredient turned murderous against a whole neighborhood. The Great Molasses Flood was a terrible Boston First; it presaged industrial accidents to come later in the twentieth century and established precedents, however slow to develop, for dealing with corporate irresponsibility, malfeasance, and greed. Rules to govern the certification of engineers were put into place, and construction regulations tightened. In the gargantuan court case that eventually ensued, some of the earliest awards for pain and suffering were granted.

Most disturbing when viewed from afar, especially in our era of celebrity whistle-blowers: the disaster could have been prevented. A watchman, lowly in employment ranking, but morally higher than any supervisors and bosses he reported to, saw evidence of leakage for months and reported it—repeatedly. Isaac Gonzales, a Puerto Rican who had worked as a seaman and laborer before taking charge of the tank, became so obsessed by worry that he left the bed he shared with his wife night after night, to check on the seepage and rumbling. Realistically, he could do nothing—the response of his superiors at United States Industrial Alcohol (USIA) had been to paint the tank brown to disguise the leaks—but feeling responsible and hoping to atone for the negligence of his bosses, Gonzales would open the spigots of the tank. He let molasses flow into Boston Harbor to try to lessen the pressure on the tank's rusty walls.

The Great Molasses Flood was in danger of becoming a historical footnote, a fluke. A tank of syrup turning lethal is a First, yes, but also so odd as to be dismissable. But in 2003 Stephen Puleo, a former reporter with an interest in history, wrote *Dark Tide*, bringing to life the events and personalities surrounding this early industrial accident, lethal as a factory fire or chemical explosion. Puleo combed through volumes of transcripts of the court trial that eventually took place. The cast of characters he isolates, describes, and follows through the drama—a tragedy in the classic, inexorable sense—could not be improved upon in a work of fiction:

Isaac Gonzales, watchman on the USIA property, a native of Puerto Rico, tried valiantly to have the defective tank repaired.

Colonel Hugh Ogden, a Boston-born lawyer, Harvard graduate, and decorated war hero, functioned as auditor over the court case.

Damon Hall, lead attorney for the plaintiffs, native of Belmont, Massachusetts, appealed to "everything that is human in mankind" in his arguments.

Charles F. Choate Jr., lead attorney for USIA, a Harvard graduate, argued that anarchists were responsible for the tank's explosion.

Arthur P. Jell, assistant treasurer for USIA, was the New York–based administrator whose negligent oversight of the Boston-based tank led to the death of twenty-one people.

In the years it would take to ascertain and ascribe blame for the tragedy, it would turn out that the molasses was not so innocent and untainted an ingredient. World War I was also related, along with the fear spread by anarchist attacks and the anti-Italian sentiments associated with Sacco and Vanzetti.

The tank should not have been where it was, so close to a densely settled neighborhood. But it was convenient and profitable for United States Industrial Alcohol. When USIA sited the tank on Commercial Street in 1915, it did so because it was just a few hundred feet from the harbor, where the molasses would come in from Cuba and the West Indies, and because there were rail lines to deliver it to East Cambridge where it was processed into alcohol. The company also knew there would be little opposition from the people of the North End. They were lowly immigrants, poor, disfranchised people who kept to themselves, especially in the climate of suspicion aroused by the anarchists, many of whom were Italian.

Over five years, a legal case was put together, an enormous civil lawsuit of 119 claims, the longest and most expensive civil suit in

Massachusetts history, with almost a year of testimony. At the trial in 1924, it was established that not only had watchman Gonzales repeatedly reported the leaks to his supervisor, but that he had taken the extraordinary step of paying a call on Arthur P. Jell, USIA's assistant treasurer. Gonzales brought with him rusty steel flakes from inside the tank. In the safety and isolation of his office, Jell dismissed the flakes and Gonzales's concerns. The ambitious administrator had no engineering background and did not consult with engineers on the tank's plans, parts, and construction. The tank was deficient in materials from the start—the gauge of steel used was too thin—and was rushed into production by Jell, who sought a promotion. The supposed test he ran and reported was a sham. To check for leaks, he only partially filled the tank—six inches of water in a fifty-foot tank—hoping to complete construction to receive a pending shipment. He was successful.

Technically, the presiding court official, Colonel Hugh Ogden, was not a judge but an auditor, in the position of evaluating arguments, evidence, and testimony to decide whether the case should proceed to full trial. But Ogden knew from the start that if he found in favor of the plaintiffs, USIA would be likely to settle, rather than to risk bringing the case to trial, where damages would be larger.

"I cannot help feeling that a proper regard for the appalling possibility of damage to persons and property contained in the tank in case of accident demanded a higher standard of care in inspection from those in authority," Ogden would write in his legally elegant fifty-one-page decision in April 1925. "I believe and find that the high primary stresses, the low factor of safety, and the secondary stresses, in combination, were responsible for the failure of this tank."

The simple brown syrup, molasses, seems so homey a thing. Brought into Boston since the early 1600s, the ingredient imparts a caramel-like sweetness to Indian pudding, Boston baked beans, and molasses-ginger cookies. It was indispensable in colonial times for flavoring salt pork. It was used to distill rum and was essential in the vastly profitable triangle trade that helped to build New England.

As every schoolchild learns, molasses, the by-product of the refinement of sugar cane, was brought by ships to New England and turned into rum. The rum was shipped to West Africa, where it was traded for slaves. The slaves, human cargo, were shipped to the West Indies and traded for still more molasses—and so on, leading to good cooking, good drinking, good profits, human misery. A dark secret of many New England families, whose descendents may well have become staunch abolitionists, is that their generations of prosperity originated in the slave trade.

Molasses fueled the American Revolution. The colonists chafed under British rule, and more so under British larceny. The Molasses Act of 1733 became the Sugar Act of 1764, affecting imports of molasses, sugar, textiles, wine, coffee, and indigo, forcing the resentful colonists to pay taxes to the British. Samuel Adams urged his compatriots to boycott British goods. James Otis decried Britain's "taxation without representation." Resentment simmered into outrage, igniting the Revolutionary War in 1774.

Molasses's protean form—and almost alchemical agency—led to the Great Molasses Flood and the twenty-one deaths, hundreds of injuries, and a broken, despairing North End community. Homey molasses was used to create industrial alcohol, a key ingredient in the munitions—high explosives and smokeless powders—needed for World War I. Huge profits were to be made, and were. According to author Stephen Puleo, more than 632 million pounds of smokeless powder was produced in the U.S. between April 1917 and November 1918 (when the Armistice was signed), equal to the combined production of England and France. The U.S. production of high explosives was 40 percent larger than England's and almost double that of France for the entire year of 1918. Little wonder that USIA did not want to impede production by making necessary repairs.

And so the tank on the Boston waterfront was filled to its two-million-gallon level seven times in 1918 alone. Through all these months, the tank leaked so obviously that neighborhood children filled buckets. Isaac Gonzales continued his frantic, middle-of-the-night visits to check on the rumbling tank's condition, watching helplessly as molasses seeped from its seams.

The case of the Great Molasses Flood receded into history. The Italian North Enders mourned their dead. Some cursed the country they had come to, became bitter, and withdrew still more. Others realized how their insularity had contributed to the disaster; if they had been citizens and been able to vote, perhaps they could have kept the tank from being built just steps from their homes, and brought concerns for their safety to elected officials. They saw how the vote could be used. Citizenship among Italian Americans increased. As the "foreigners" became less foreign—more involved in the larger community—prejudice decreased. The workers injured in the flood, including firefighters, were devastated. Most never recovered (and this was before modern painkillers, Social Security, workman's compensation). But the outrage of those who survived led to greater concerns about public safety, more vigilance over Big Business, and tightened construction regulation.

The primacy of molasses—three hundred years of it in Boston and New England—passed. World War I came to a close, and with it the inflated need for industrial alcohol. Prohibition arrived, canceling the commercial manufacture of rum. Sugar—neat, white, pure—became America's sweetener. The crash of 1929 ended America's love affair with Big Business, though not indefinitely.

Later in the twentieth century, other versions of the Great Molasses Flood would play out—environmental damage in canals of toxic waste, in lakes of poisoned and poisonous fish, in polluted air that led to asthma. Other heroes and righteous men such as Isaac Gonzales and Colonel Hugh Ogden would arise and act. Sometimes they would be heard, and sometimes not.

COMMUNICATION

NEWSPAPER

I t was the *People* magazine of its day. *Publick Occurrences, Both Foreign and Domestik* answered the colonists' need for news about their own lives and also dished the dirt. It was snapped up like hotcakes, but so enraged the censors that they quashed it after a single coveted edition. But still, *Publick Occurrences*—published September 25, 1690—was the first American newspaper. Its publisher, Benjamin Harris, planned it as a monthly, or, to put the prospectus in his own well-chosen words, the paper was to be "furnished once a moneth (or if any Glut of Occurrences happen, oftener)."

The tidy mini-tabloid is recognizably, delightfully, a newspaper. It makes a great read, even now. Think of the columns in your local newspaper, where goings-on are catalogued one after another —news of people and events you chat about with neighbors over coffee. In the single precious edition of *Publick Occurrences*, you would have read similar accountings, along with a write-up of the first Thanksgiving, a suicide in Watertown (the deceased—a recently widowed man suffering from "the Melancholy"—was found "dead with his feet near touching the ground"), a fire in Boston, atrocities committed by "barbarous Indians," and local public health threats (Smallpox, Agues, "Epidemical Fevers"). News of the day had been observed, reported on, written, and printed by Benjamin Harris, arguably America's first journalist. He lived in Boston just eight years, but arrived well skilled at agitation and publication.

Benjamin Harris (1673–1716) was a known troublemaker, which

is how he wound up in Boston in the first place, where he continued to make headlines. During the seventeenth century, printers were often editorialists, involved in all aspects of the ink trade. In London during the 1670s, Harris printed the works of others—books, pamphlets, broadsides—as well as his own work.

He was virulently anti-Catholic and anti-Quaker, and attacked these faiths in his publications. The ruling royals and the nobs were unconcerned until Harris began to opine on their misdeeds. From 1679 to 1681 he published *Domestick Intelligence or News from City and Country*, featuring local news, including the sensational. In 1679 he published *Appeal from the Country to the City*, lambasting the king and Parliament. He was promptly charged with sedition, fined heavily, and sent into the murk of King's Bench Prison in London. In December 1680, friends bailed him out and paid his fines—and by the following April, he was at it again. The inveterate pamphleteer published yet another edition of *Domestick Intelligence*, overflowing with opinion. In 1685 he composed, printed, and distributed *English Liberties*, an assault upon Catholics—this during the reign of James II, a Catholic. Another stay at dank King's Bench Prison loomed.

Harris and his son hightailed it out of London and set sail for Boston, in search of greater freedom. For a brief time he found it—the story is worthy of a ballad—and created a small, progressive community, a haven for people, including women, who loved conversation, coffee, and books. Author Peter F. Stevens describes this haven in *Notorious and Notable New Englanders*, in a fascinating chapter, "New England's First News Hounds." If ever there was such a hound, it was Harris, who arrived on these shores with his son, both schlepping crates of books, in the autumn of 1686. Undaunted by the crossing, Harris soon set up shop on the corner of High and Great Streets (Washington and State Streets today), and his bookstore/print shop became a hangout. In 1690 (and we thought we invented bookstore cafés), the lively printer/bookseller started serving the classic accompaniments to literature—coffee and tea, the beverages of stimulation and fellowship—in his inviting shop, the London Coffee House.

Author Peter Stevens paints a pretty picture of the café: Cotton Mather hobnobbing with tinkers and tradesmen, and "Puritan housewives in their bonnets, petticoats and capes" reaching for the work of Anne Bradstreet (1612–1672), the poet, wife of colonial governor Simon Bradstreet, and mother of eight. In her writings, Bradstreet contemplated her surprising settlement in the New World, as did her readers in the London Coffee House.

As a rule, women in seventeenth-century Boston were not permitted to enter taverns and inns. Imagine then, the newfound pleasure and exhilaration of turning from slopping pigs on a summer morning to reading poetry in the afternoon; of cradling a cup of tea in both hands, in the very dead of winter, as opposed to wringing out stiff clothing in ice water.

Ensconced in Boston, Harris returned to publishing, including the well-used *Tulley's Almanac* and *The New-England Primer,* the classic textbook for schoolchildren, which dominated the market for over one hundred years. But journalist Harris was a keen observer, itching to publish a newspaper again.

On Thursday, September 25, 1690, *Publick Occurrences, both Foreign and Domestick,* the first and only edition, emerged wet and glorious from his press. A small but enthusiastic crowd gathered outside the shop, waiting to receive and hold the inky, handsome newspaper in outstretched arms, to read of events in their own community, documented, described, set up for debate.

The first American newspaper is a delight to behold. (The Massachusetts Historical Society has a facsimile copy made from the single surviving original in Kew, England.) It looks as we would want it to: a bookletlike tabloid the size of a magazine (about 7.5 by 12.5 inches), made of two folded sheets printed on three sides, with two columns of neat newsprint on each page, and the last page left blank for readers' notes—successive readers' notes, actually, as it was expected that the newspaper would be passed on.

The newborn publication was easy to hold in one's hands or to spread upon a table; it fit well on a colonial knee and in an aproned lap. The contents were delicious, with or without coffee: local news, lively writing, and gossip to beat the band. Colonists read Harris's newspaper while drinking coffee and tea. They read portions aloud to each other, and discussed the issues of the day. They debated, marveled, snickered. A new era beckoned.

Alas, the Puritan establishment found portions of *Occurrences* unfit to print. Harris's reports of atrocities committed by American Indians aligned with the British were politically incorrect. His descriptions of suicide and murder were lurid. Naughty bits about the king of France—attributing difficulties with his son to familial adultery ("if reports be true, that the Father used to lie with the Sons Wife") were definitely out. Boston Puritans did not regard this type of reportage as proper reading material, and declared their "high Resentment and Disallowance of said Pamphlet."

Governor Simon Bradstreet forbade the distribution of the newspaper. (Yes, this is the husband of the aforementioned poet Anne Bradstreet, whose work Benjamin Harris lauded, promoted, and sold in his bookstore-café.) Chief Justice Samuel Sewall, whom Peter Stevens describes as a "heavily jowled jurist," buttressed the governor's injunction by citing the unlicensed status of the newspaper. Printer/publisher/journalist Harris, who should have lived in another century—and who in some ways does—was muzzled once again. He remained in Boston a few more years, accepting other printing assignments, none of them much fun, and in 1695 returned to London, where he busied himself with profitable publishing ventures.

The coffeehouse community he helped to create—near today's Downtown Crossing, and not far from where Newspaper Row would someday exist—had no similar place to gather. Its members did not stop drinking coffee, or reading, or speaking with one another, but the ease and vitality of association promised by the coffeehouse-bookstore-cafe community was gone. Nothing replaced it.

Other newspapers followed. They had less flavor than *Publick Occurrences*. John Campbell, a Scottish émigré, published the *Boston News-Letter* from 1704 to 1722, avoiding censure by mainly covering news from abroad. Anything likely to offend the colonial governors and clergymen, he removed. It would be many a year and many a battle till freedom of the press tolled its bell. But the people at Benjamin Harris's coffeehouse had been awakened—to the import of printer's ink, good coffee, news, views, and exchange—and found that a newspaper was not only necessary to just governance, but to just plain fun.

They had glimpsed something new to them—the idea of a regular appearance of a lively, irreverent, local newspaper, and coffee and books in a cozy place, and men and women reading together in public. It would take a few generations for the liveliness of *Publick Occurrences* to reappear in American journalism. It would be more than two centuries before men and women shopping for books would sit down at a table and drink coffee together with ease, and more time still before the appearance of that tableau of American culture: working people at a lunch counter, drinking coffee together, reading the newspaper.

NOVEL

I t was a scandal within a scandal, delicately wrapped in piety and propriety, and then, as though radioactive and needing multiple shields, triple-wrapped in mystery and secrecy. *The Power of Sympathy,* the first American novel—the story of a quasi-incestuous, adulterous relationship, and the suicide of a previously sheltered young woman—was fiction based on an actual, public scandal. Published anonymously, it was attributed to a woman author, but "she" was actually a man. It was passed off as a moral lesson to young ladies, but was a tale of sexual power and abuse. Though sought by readers and initially sold at Boston's best bookstores, including Isaiah Thomas's shop at 45 Newbury Street, in later reports it seemed to disappear from bookstore shelves.

For many years, it was thought to have been suppressed, though this charge now seems unfounded. While nothing to sneeze at as a literary accomplishment (what's not to be proud of in writing the first American novel?), *The Power of Sympathy* may be one of those projects—familiar to skilled cooks, research scientists, veteran legislators, and, yes, novelists—whose raw materials, assembly, and process are ultimately more compelling than the final product.

The novel form, which we today take for granted, is a fairly recent development in publishing. (Scholars generally place the "invention" of the English novel in the eighteenth century, arising from genres in both France and England, as well as ancient Greece.) Though newspapers had already been published in the

U.S. (see page 43), the art of fiction needed a few years after the Revolutionary War to develop. Perhaps nothing fictional was as dramatic as the process of separating from the British. Novels, like plays, were also viewed with some suspicion by Boston Puritans. It is no coincidence that the first novel was packaged as a moral lesson. It is "Dedicated to the Young Ladies of America," according to an advertisement of the day.

In *The Power of Sympathy*, published in Boston in 1789 by ubiquitous printer Isaiah Thomas—one is tempted to call him Printer to the Stars—an unseemly seduction takes place between a man and a woman. The woman, called Ophelia, is the unmarried sister of the man's wife. She—not, of course, the wily, self-important seducer—kills herself. The sorry tale happens in an upper-class family, in beautifully appointed rooms. The story is "Masterpiece Theatre" material.

In real life, this was a widely known drama played out in 1780s Boston. The beautiful, intelligent, educated daughter of one of Boston's best families, Sarah Wentworth Apthorp, marries Boston's Prince Charming, Perez Morton, a lawyer. Well into their marriage, her philandering husband initiates an affair with his sister-in-law, Frances, called Fanny. They openly carry on, and privately as well, and pregnant Fanny leaves the censorial gaze of Boston society to have their child. Shame and ruination is brought upon the family. Fanny Apthorp poisons herself.

To summarize, then, the main characters in the real-life tragedy are:

Sarah Wentworth Apthorp became Mrs. Perez Morton and published poetry under the name Philenia.

Perez Morton, a leading Bostonian, supposedly the right stuff, though a philanderer, seduced his sister-in-law, Fanny, and later, unsuccessfully, tried to suppress distribution of a novel about their affair.

Frances (Fanny) Apthorp, visited her married sister Sarah, fell in love with Perez, and became pregnant with his child.

William Hill Brown, a kindly, literate fellow, a writer and neighbor of the Apthorp and Morton families, published widely in his time.

In status-conscious, gossipy, eighteenth-century Boston, houses were also important "characters," functioning as seats of power. Sarah and Fanny, "the Apthorp sisters," grew up in a fine home on State Street. Their wealthy parents were friendly with the leading lights of the day, including John and Abigail Adams. Sarah and Perez Morton, a power couple—she a published poet, he a lawyer, speaker of the Massachusetts House of Representatives, and state attorney general—resided in a grand house at the junction of Dudley and Alexander Streets in Dorchester. Their baronial mansion —a center-entrance colonial with columns, portico, pediment, and deep side porches, all on a graceful rise of land—was the talk of the town. It commingles many Firsts: The house was designed by Charles Bulfinch (see page 105), Sarah Morton's first cousin. Around the same time, Sarah's portrait—dewy complexion and fair features captured—was painted by Gilbert Stuart, who in his off-hours sketched the famous editorial cartoon "Gerry Mander" (see page 207). And though the house was torn down toward the end of the nineteenth century, it long graced Dudley Street, the location of a contemporary urban village (see page 187) with houses that echo Bulfinch row homes.

The nicely produced first novel—complete with a fancy copper-plate frontispiece portraying "Ophelia," the suicide based on Fanny Apthorp—was marketed much as a book of its type would be today. Though it is an epistolary novel, following letters written and received by various individuals, and has numerous plots and subplots, the juiciest—the one based on the public scandal—was promoted. And yet it is a tiny portion of the two-book novel! Writing about *The Power of Sympathy* in a textbook on early American literature, Richard Walser quotes an article unearthed from the eighteenth-century newspaper, *Herald of Freedom*, published on January 16, 1789. The sotto voce tone of the review suggests the presence of a press agent, and a good one:

> We learn that there is now in the Press in this town a Novel, dedicated to the young ladies, which is intended to enforce attention to female education, and to represent the fatal consequences of Seduction. We are informed that one of the incidents upon which the Novel is founded, is drawn from a late unhappy suicide.

Though the novel was published anonymously, it was widely assumed that Sarah Apthorp Morton, aka the poet Philenia, sister of Fanny, wife of seducer Perez, was the book's author. (Shades of Nora Ephron writing a thinly disguised novel of her husband's infidelity.) It was plausible—she was a published writer, the events described played out before her eyes, in the very chambers and corridors of her home, and who could blame her for framing the story of the family tragedy as a moral lesson, not to mention transforming humiliation into art? For 105 years, until revelations were made by literary scholars in 1894, *The Power of Sympathy* was attributed to Sarah, but the book was actually written by Morton's neighbor, William Hill Brown. The young author was close at hand during the scandalous seduction, watching the carriages come and go and hearing the hysterical weeping from the neighboring house.

After considerable brouhaha and early attention, the novel, and the scandal, passed from public concern. *The Power of Sympathy* was reprinted about a century later, and there was a brief flurry of interest. Its author, William Hill Brown, just twenty-four when the novel—his first and America's first—appeared, continued to publish. One theory as to why the book didn't do better was that the scandalous story promoted as the novel's "hook" was only a small element of its several plots. And William Hill Brown could not call attention to his authorship. He published anonymously to protect himself from the wrath of powerful attorney and politician Perez Morton.

Poor Fanny Apthorp died, and the life of the child fathered by her brother-in-law, Perez—assuming the child survived—is lost to our knowledge. Would that he or she had written a book.

Sarah Wentworth Apthorp Morton—the wronged wife who "got over it," and who continued to publish poetry as Philenia— seems to have fared best of all. She continued to live in her grand Dorchester house, was respected for her poise and strength of character, and outlived her troublemaking rake of a husband by a decade. Almost till the end, Sarah wrote verse in the "sky parlor" of her Dudley Street mansion. She died at age 86 in 1846, having learned the power of sympathy, and those of understanding, forgiveness, and fortitude.

In any age, a woman with poetic insight and independent income, a blessed combination, can persevere and triumph through much.

THOMAS ALVA EDISON AND THE ELECTRIC VOTE RECORDER

❦ 1 8 6 8 ❧

L ike many an ambitious young person in Boston, experimenting in life and work, Thomas Alva Edison (1847–1931) started his trajectory here, then cut loose to seek his fortune. In 1868, Edison invented and patented his first invention in Boston: an electric vote recorder for legislative use, the "Electrograpic Vote Recorder and Registrar."

It was a compact, ingenious, small machine, a bundle of wires and metal mounted on wood. It involved politics—that Boston specialty—and electricity, the elixir of Edison's day, as genes and chips are of our own. The machine was designed to tally and transmit the votes of legislators. It wasn't successful—which is to say, it didn't sell—but it was successful in teaching Edison how to "expend time."

Thirty years later, in 1898—by which time Edison had become the eccentric inventor celeb par excellence—the great American novelist Theodore Dreiser interviewed him for the magazine *Success*. Theresa M. Collins quotes some juicy excerpts in her book *Thomas Edison and Modern America*. Dreiser, himself a celebrity, asks the wealthy inventor about the Boston experience and his first invention: "Yes, it was an ingenious thing," remembers Edison, who has just edged into the room wearing well-worn workmen's clothing. "Votes were clearly pointed and shown on a roll of paper, by a small machine attached to the desk of each member. I was made to learn that such an innovation was out of the question, but it taught me something.... To be sure of the practical need of, and demand for, a machine before expending time and energy on it."

Still, the young Edison hadn't suffered in Boston. He had the time of his life.

He had started his trajectory in Milan, Ohio, a busy wheat port. In an almost Biblical bread-cast-upon-the-waters event, as a boy growing up near railroads, the quick-witted Edison had rescued a small boy from the path of a runaway boxcar. In gratitude, the boy's father—wouldn't you know it, a telegraph operator!—offered to teach him the trade. Edison was a natural and started working in a local office. Even before training, he had been intrigued by the device and worshipped Michael Faraday (1791–1867), the English scientist who studied magnetism and electricity. The teenaged Edison was able, affable, and winning, and easily found jobs. He also kept losing jobs, as he was not the nose-to-the-grindstone type, and couldn't keep his hands off the other equipment and machinery. He examined, took apart, and experimented with everything he could get his hands on, including what didn't belong to him and had nothing to do with his job.

Edison decided to head for Boston, a bastion of civility, intellectual activity, and avidity for the telegraph. What a place was Boston for a telegraph-techie in 1868! Edison was only here for a year, living near the gold-domed State House, but he was in the right place at the right time. He mixed it up with other clever young men, most of whom were also biding their time at the local telegraph office, which functioned as a kind of garage band for inventors.

Telegraphy was the place to be for inventors interested in communication and electricity, and Boston was Telegraphy Central. While it was second in "go-go" atmosphere to New York City, the Boston telegraphy scene was well established, intellectually sophisticated, and research-friendly. Western Union had four offices here, and Franklin Telegraph Company, a rival to Western Electric, was also based in Boston. The city had the first fire-alarm telegraph system in the nation. Edison, who had lots of experience taking press wire copy (which he routinely read, educating himself), got a job at Western Union, and spent his off-hours hooking up with other telegraphy buffs and writing articles for a publication called *The Telegrapher*. Soon he was renting space at a legendary machine shop run by manufacturer Charles Williams Jr., where Alexander Graham Bell (see page 59) would also work. (Bell's future machinist, Thomas Watson, was already working at Williams's shop when Edison signed on.)

It was a free-form atmosphere—lots of high-spirited, brainy young men, ideas aplenty, technical journals and newsletters tossed back and forth, and inventions quickly rushed into production, advertised, and adopted. Descriptions of these inventor-entrepreneurs make them sound a lot like the early computer wunderkinds, as do their seat-of-the-pants ventures and tryouts of new products. Once in business in Boston, having left Western Union, Edison sold private-line telegraph service to firms who wanted communication between their salesrooms and factories. Installation was comically casual! Edison and his cohorts installed wires outside connecting buildings, using the roofs of houses and offices, never paying a penny for use. They just showed up, bright and bushy-tailed, sounding official. "It never occurred to me to ask permission from the owners; all we did was to go to the store etc and say we were telegraph men and wanted to go up to the wires on their roof and permission was always granted," says Edison in Paul Israel's *Edison: A Life of Invention*.

During this protean year, Edison was always working on several inventions at once, including "printing telegraphs" for stockbrokers, facsimile telegraphy (transmitting handwriting, diagrams, images—a forerunner of the fax), and applications of "double telegraphy" (running two messages simultaneously in opposite directions on one wire). This pattern of simultaneous working through of ideas would characterize and dominate his life.

In 1870, after a spate of difficulty finding backers, Edison moved to New York City, the indisputable capital of finance, commerce, and the telegraph industry. By the early 1870s he was working in Newark, New Jersey, gaining an international reputation in telegraphy and a slew of manufacturing and research contacts. He never went back to Boston because he had found its backers too tight. "N.Y. appears to be quite different to Boston," he wrote in a letter that author Theresa Collins refers to. "People here come and buy without your soliciting."

The reason that Edison's first patent, the electric voting machine (in Boston, 1868), was unsuccessful is that its speed was inappropriate for the Bostonian legislative process or atmosphere. On Beacon Hill, then as now, a crisp yes or no vote was rare. Instead, legislators used the roll call as stage time—to blather, palaver, jawbone, declaim, and generally run their mouths. Their time-consuming grandstanding was at odds with the speed of the new machine, which could only transmit "yes" or "no." But never mind. Thomas Alva Edison learned what he needed to in Boston— "to be sure of practical need and demand"—and went on to give the world what it wanted: the incandescent lamp, the phonograph, the microphone, and more. He patented over one thousand inventions.

The connections he made in Boston and the lessons he learned spurred him on, as did the companionship, collegiality, and abundant fun. Like many a young person in Boston, Thomas Alva Edison figured himself out here, and went forth.

TELEPHONE

❦ 1 8 7 6 ❦

"Honey, I'm home!" and "I'm leaving the office now" are among the most frequently spoken sentences in America. The latter is generally spoken over the telephone and has been for 130 years. The initial utterance of this everyday update was made in Boston on the "harmonic telegraph," soon to be known as the telephone. Wires carried an individual human voice from a husband in Boston to a wife in Somerville. He spoke, she heard, she spoke back, they conversed. It seemed like a magic trick.

The trick carried a particular voice—unique in accent, pitch, rhythm, resonance, timbre, inflection, pauses between words and thoughts, and the personal language and references—between a man and a woman across miles. A man in his office talking to his wife in their home!

The year was 1877. The caller was Charles Williams, the spark plug to so many inventors and inventions, proprietor of the Charles Williams Electrical Supply House, the legendary machine shop at 109 Court Street, Boston. Williams was calling Mrs. Williams, who by that time in her life thought she has seen and heard everything. But she had not. Here was a direct, live connection between two people, not merely printed words on a yellow telegraph form arriving in due time. The device was wired into a home, as opposed to a business or a doctor's office, like the earliest telephones; it was the first private residential phone line.

The harmonic telegraph had been invented (accidentally) the

previous year by the learned, scholarly, philosophical educator Alexander Graham Bell, assisted by the brusque, vulgar, and altogether essential machinist Thomas Watson.

Bell, whose "day job" was to teach the deaf to communicate, had been working with Watson, an expert in electricity, for several years to try to find a way to transmit sound across wires, building on the work of earlier telegraphers. On June 2, 1875, a reed that Watson had attached too tightly to an electromagnet froze in place; Watson dexterously plucked it loose. Bell, who happened to be at the end of a receiving reed in the next room, heard the resonant twang. Only a man trained in the fine points of sound and elocution would have recognized this tone with overtones, the transmitted "pluck" in the next room. The reed had transmitted an undulating current that could vary in intensity as air varied in density when a particular sound passed through—a pivotal discovery.

Bell and Watson continued to test their machine. "Mr. Watson, come here, I want you," said Bell, on March 10, 1876. (He had already applied for a patent.) An instant later, in rapt response, Watson came flying into the room. Bell's exclamation to Watson, supposedly uttered because he'd spilled battery acid, is a famous line of dialogue—we have been reading it in books since grammar school—announcing the first transmission of speech. But Charles Williams's less famous Boston-to-Somerville chat with his wife was homey, human, a particular aspect of the harmonic telegraph —soon to be renamed the telephone—that particularly pleased Bell. He had sought for years to develop a form of speech that would enable deaf people to communicate with the hearing world and with each other, and he also worked at the end of many workdays on a more abstract problem: the transfer of multiple sounds on a wire. He was an educator who became an inventor to promulgate his ideas, including the idea that sound was amazing and beautiful (he was an accomplished musician), and that people should be able to speak, to share their inner worlds, with each other.

Bell loved, appreciated, and was fascinated by sound—all of it: music, voices, noise, and words. (His distinctive way of answering the telephone would become "Hoy!") Born in Scotland, an émigré to Ontario, Canada, as a young man, he grew up in a family of educators. His grandfather, father, and uncle were noted elocutionists; his father developed Visible Speech, a system of teaching deaf people to speak by training them in the anatomy and physiology of speech, so they could learn to produce sounds even if they couldn't hear them. Bell's mother, whom he adored, was deaf, and Bell was deeply sensitive to her isolation. Like a child who grows up with illness in the family and becomes a doctor—not only to help, but because he is familiar with a particular world—Bell felt it was his mission to bring speech to the world of the deaf, so that they could be brought into the larger world, and so that the larger world could appreciate them. From boyhood on, he was an unusual combination: artistic and romantic, scientific and practical.

Bell migrated from Brantford, Ontario, to Boston in 1871, when he was twenty-five years old, sent by his father to teach Visible Speech at the Boston School for the Deaf. He entered Victorian Boston as though it were a stage set designed especially for him— a place of parks and British-style buildings, the comforts of armchairs and cozy tobacco shops, the stimulation and validation of progressive schools for the deaf, and educators who supported his methodology. Young Bell found pals, meals, and boarding houses. He would soon find and make use of the lively lecture scene at Harvard and M.I.T., and the wonders of invention at Charles Williams's Electrical Supply House, where agile-minded inventors found agile-fingered machinists who could translate their ideas into objects.

Bell continued to teach at the School for the Deaf during the day, as his experiments in the evenings progressed. In 1872 he opened his own School of Vocal Physiology and Mechanics of Speech (the "oral" method as opposed to sign language), began to teach at Boston University, and made the fortuitous acquaintance of Gardiner Greene Hubbard, a wealthy Cambridge lawyer with whom Bell formed a patent association in 1873. He also met Hubbard's young daughter, Mabel, who was deaf. Mabel Hubbard would first become Bell's student, and later on his wife.

Using the Visible Speech method his father had pioneered, Bell slowly, patiently taught sixteen-year-old Mabel to speak. He worked with her for many months, and though he was proper and shy, eventually told her that she had a lovely voice, which he himself had evoked. Mabel had not heard her own voice since childhood, when scarlet fever had caused her deafness. She never heard the voice of her teacher, Bell, but responded to him. They would forever share this distinctive engagement and bond—the way they had, like all lovers, evoked hidden voices in each other, though in their particular case it was literal.

Following the astonishing events in the laboratory in June 1875 —the historic twang or pluck—Bell applied for a patent on his harmonic telegraph on February 14, 1876, beating out inventor Elisha Gray by just three hours. A patent was issued on March 7, 1876— three days before Bell first spoke to his assistant via his invention. Bell and Watson continued to work on their device and to raise money from investors. They even arranged road-shows—public demonstrations of harmonic telegraphy—including a most enthusiastic, well-attended hookup between the cities of Salem and Boston.

As with the early use of ether (see page 81), other inventors had run neck-and-neck with Bell, developing their own solutions— not only Elisha Gray, but also Thomas Alva Edison and Amos E. Dolbear. There would be six hundred lawsuits over the years, exhausting Bell with court appearances and testimonies, which he nevertheless became quite good at, with his imposing demeanor, impressive knowledge, and perfect elocution.

Bell gave great gifts to Boston and the nation. In just thirty-nine years after his invention, there would be eleven million telephones in the U.S., revolutionizing communication. But Boston gave gifts to Bell, as well: a welcoming educational environment; encouragement of his teaching of the deaf; the charmed, exhilarating world of Charles Williams's shop; and the vigor and talent of his assistant, Thomas Watson, who became a lifelong friend. He met canny, prescient financial backers who enabled him to become wealthy, and he found his wife, Mabel, and their special connection.

When they married in 1877, Bell gave Mabel Hubbard two gifts. One was romantic and spiritual, the other practical and down to earth. He presented his young bride with a glowing cross, set with eleven pearls, and 1,497 shares in Bell Telephone Company stock.

Hoy!

TV FOR THE DEAF
AND THE BLIND

✎ 1 9 7 2 , 1 9 9 0 ☙

Since the dawn of the television age, deaf people could watch TV, but they could not hear it. No news shows, no sitcoms, no how-tos, no talk shows, no late-night comedy, or must-watch trash TV.

The talk about TV that pervades our culture—discussion about events on the news, participation in fictional TV families, renditions of characters and skits—were not part of the world of the deaf.

Without access to the serious stuff—news, documentaries, commentary—deaf people were denied an important part of being a citizen, an understanding of events. Without access to the fun stuff —sitcoms, serials, comedy—they were denied participation in popular culture. Consider, well into the television age, deaf people had no access to Charlie Brown at Christmas, play-by-play of sports, soap operas, Miss America, or Mary Tyler Moore. No inauguration of presidents, including John F. Kennedy, hatless and coatless in the wind. Not understanding TV kept deaf people out of the loop, increasing the isolation they already experienced. The millions of "TV-deprived" were not only the congenitally deaf and people who had lost their hearing because of accidents or illness, but elderly Americans suffering common, progressive hearing loss and the resulting disconnection.

Their disconnection—social, political, cultural, psychological—has dramatically improved. Deaf people now receive TV. Their access began in Boston.

In 1972, WGBH-TV, Boston's public television station, broadcast the first television program captioned for the deaf: *The French Chef.* (If you've never seen a captioned TV program, think foreign films with subtitles, although captions for the hearing-impaired convey not only dialogue, but other sounds, as well.) Today, almost every deaf person in America has access to captioned TV—thanks to the staff of WGBH Boston and a visionary head of a federal agency who was himself deaf.

Captioning happens—words appear on the TV screen—because of technology. But the story of captioning is about people—people who wanted to be in the thick of life but who could not hear, and communication professionals who wanted to extend the reach of television.

One of the people in the story is Dr. Malcolm Norwood of the Department of Health, Education, and Welfare, considered "the father of captioning"—an educator, activist, and pioneer in captioning film and TV. Dr. Norwood (1927–1989), affectionately known as "Mac," was deaf. He knew personally and professionally what hearing-impaired people missed. He and his wife, who was also deaf, raised "hearing" children. Every day their children watched TV that neither parent understood.

Another principal player: Phil Collyer, the modest, affable producer/director who set up The Caption Center at WGBH-TV and was its first director. Collyer, now sixty-six, went on to do lots of other work in television, including producing and directing sports programming, directing operations for WGBH, and producing the station's annual fundraising auction, but he considers the captioning project "the most rewarding part" of his career. He hired a small staff who became captioning pioneers. Their work delivered television to a new audience: millions of deaf Americans.

Dr. Norwood had already involved the federal government in a captioning project—preparing feature films for use in "deaf clubs," social organizations for the hearing-impaired. "It wasn't publicized, but you could go into one of these clubs and see a first-run movie with captions," remembers Phil Collyer. For captioning these films, Norwood had turned to film experts. When he got the idea of captioning TV, he turned to TV pros. He contacted WGBH, a pioneer in educational television, asking if they would be interested.

Collyer, barely in his thirties and with no experience in special education, was intrigued, and led the efforts to create "open captions"—a version of an existing program with superimposed, expository subtitles. He hired staff, including deaf staffers, and trained them. Collyer cites these young men and women—now middle-aged—as part of what was so rewarding about setting up The Caption Center. Many of these staffers have become lifelong captioning professionals.

Soon, Julia Child became part of the mix. WGBH producers presented *The French Chef* as the test broadcast, reasoning that no one could resist Julia. Julia with captions was a smash. In 1972, at the urging of Dr. Norwood—ever encouraging, ever smoothing the way—the U.S. Department of Health, Education, and Welfare funded the captioning of *The French Chef*, the first nationally broadcast captioned TV show. Within weeks, deaf viewers were chopping mushrooms, simmering coq au vin, turning omelettes out of pans.

TV for the deaf began in the atmosphere of the disability rights movement, which evolved from the civil rights movement, in an era focused on equal opportunities and access for all people. In 1973, Congress passed the precedential, widely influential Rehabilitation Act. Its many provisions included forbidding employers with federal contracts from discriminating against disabled individuals in the hiring process, and mandating affirmative action programs for disabled people. Within a few years, the public landscape was transformed: Ramps were placed on public buildings. "Curb cuts" to accommodate people in wheelchairs were made in the sidewalks of New York City, chiseled into the granite curbs of small New England towns, molded into the "ce-ment" of Southern suburbs. Sign language interpreters stood next to speechmakers. Actors in wheelchairs appeared on stage—when being handicapped was not part of the play. Boston, with its history of education for the deaf and blind communities, was particularly progressive. In 1972—the year captioned TV began—the Boston Center for Independent Living was founded. Its mandate: extending civil rights to individuals with disabilities.

Collyer, a tinkerer, grew antsy with "just" captioning Julia. "I remember looking at it one day and thinking, it's Julia, it's one person. It's easy to tell who's talking. What if we were working with two people, or six people? It was a stunning moment." Soon Collyer and his staff devised techniques and technologies for captioning programs with multiple speakers—such as moving captions to different parts of the screen to indicate who was speaking.

Still, there was no "live" captioning. Dr. Norwood had always wanted broadcast news for the deaf—to bring deaf Americans into the political process. But it seemed impossible; captioning would have to be done "same day" for the news to be relevant.

Back at WGBH, Phil Collyer pushed to caption a presidential inauguration. It had never been done. In January 1973, newshounds and politicos predicted a dramatic announcement by President Richard Nixon during his second inaugural address. "We expected Nixon to announce the end of the Vietnam War," remembers Collyer. President Nixon did not announce an end to the Vietnam War. But his inauguration was captioned and broadcast. For the first time, deaf people watched and understood the televised inauguration of an American president.

Nixon's speech was twenty-two minutes long. It took Collyer's captioning pros four hours to caption it; the speech aired at 6:00 p.m. the same day. Collyer realized that twenty-two minutes was the same length as the nightly newscast, minus commercials; if he could set up a system whereby a team of captioners was assigned segments of an early feed of a newscast, a newscast for deaf viewers could be broadcast the same day.

In December of 1973, WGBH went on air with a nightly version of *The Captioned ABC News*. Staffers took the Boston feed from the ABC line at the Prudential Tower at 6:00 p.m. (it aired locally at 6:30 p.m.), broke the staff into work units—each troop assigned a series of production tasks, and captioning like crazy—and went on air five days a week at 11:00 p.m. Captioning the nightly news was a tour de force, a combination of TV production precision, ensemble performance, almost literal choreography (requiring a young, fleet staff!), and esprit de corps. When *The Captioned ABC News* went out in New England, and then along the Eastern seaboard, deaf people all over America petitioned their local public television stations to carry it. They soon did.

Captioning changed over the years. In the beginning, programs were "open captioned." Captions appeared only on specially prepared programs that were rebroadcast to the deaf community. Later, it became possible to "encode" captioning information in the TV signal such that any TV (equipped with certain technology) could receive transcriptions. During the early 1990s, it became mandatory for this feature to be in every TV sold in the U.S.

Today, virtually all television programming is "close captioned," built in before a show is broadcast. "Captioning has become a cottage industry," says Collyer. There are captioning centers all over the U.S.; the Media Access Group at WGBH continues to be a major player, captioning both public television and commercial programs. Captioning is used beyond serving the deaf, to teach reading, to educate children with learning disabilities, to assist those learning English as a second language, and to serve bilingual audiences—to caption presidential inaugurations in Spanish, for example.

Blind and visually impaired people also demanded access. They could hear the racy dialogue between characters on *American Playhouse*, but missed the sexy entrances and steamy glances. The vivid appearance of animals in nature shows was unavailable; the suspenseful, stealthy approach of predators invisible.

During the mid-1980s, WGBH began to research the possibilities of "describing" TV for blind viewers, providing additional audio material to compensate for what could not be seen. The station approached Washington Ear, a nonprofit description group that prepared audio material to make theater and museum exhibits accessible to blind people. In 1990, WGBH launched the Descriptive Video Service, and later that year, an entire season of *American Playhouse* was broadcast with "video description"—the first TV broadcast for the blind and visually impaired. (The same year, the Americans with Disabilities Act, the first comprehensive civil rights law for people with disabilities, was signed into law.) WGBH "describers" create special scripts, a kind of enriched narration, to accompany existing programs. These words convey visual elements such as settings, actions, body language, costumes, facial expressions, and historical period. The segments are read by a narrator to augment the presentation and fill in where visual information is eclipsed for blind audiences. No part of the original program is removed; the description is layered onto the pauses of the original program. Like captioned programs, described programs are also available on ordinary TVs, carried on a portion of the stereo television feature.

In 1993, President Bill Clinton's inauguration was the first live national event to be broadcast fully accessible to people who are deaf and hearing impaired, and people who are blind and visually impaired. This live historical event—television for all people—was produced by The Media Access Group at WGBH-Boston.

During the early 1970s, when the first *Captioned ABC News* programs went on air, a man in the backwoods of Maine received a broadcast. He lived alone. Until he saw the news of the world on TV—and the special features for deaf viewers created by The Caption Center, which popped up during commercial breaks—he had not known there were other deaf people in the world. He wrote a letter describing the experience as the best day of his life. It seemed a miracle: not only a program created especially for him, but his discovery and realization that there were others like him—many others, he concluded, to justify the cost of the program.

Captioned television was created for the deaf man alone in the woods in Maine, who grew to feel more connected. It was created for millions of other viewers, too—blind and deaf people, and hearing and sighted families with blind and deaf members, all watching together. They became connected.

SCIENCE & ENGINEERING

SMALLPOX INOCULATION

ᔥ 1 7 1 2 ᔥ

When modern people learn about smallpox—its ravages on the individual, his intimates, and community—they think of AIDS, and polio, and the politics of abortion. The dreaded but inescapable "distemper," as it was called, was horrible, disfiguring, wasting. It seemed to spread through contact with others who were infected, though germ theory was not yet known. Pasteur's conclusive research would not take place for over 150 years.

In the early 1700s, even once people knew how to control smallpox with inoculation, most Boston Puritans forbade the technique as an article of faith: They believed that inoculation—moving the pus of an infected patient to a healthy person to enable him to develop immunity—was "playing God," and that the newly infected person would not only die, but spread his illness to the larger community, a particular sin in Puritan thinking. To speak in favor of inoculation, much less to commit the act as a physician, or to receive inoculation as a patient, was to endanger the community. In the early 1720s, opposition to inoculation created a frenzy in Boston. Paradoxically, outraged Puritans committed acts of violence and vandalism to prevent a practice that would not only save individual lives, but prevent epidemics.

The man most responsible for the first use of inoculation in the New World was Cotton Mather (1663–1728). He was a Puritan and a minister, son of Increase Mather, and one of the most influential theologians of his day. Mather convinced Zabdiel Boylston, a local doctor, to try the controversial procedure, persuading him with scientific papers and eloquent personal letters. In forging this alliance, Mather brought censure and violence upon Boylston and himself. He also saved hundreds of lives in their community during the epidemic of 1721, and thousands of lives as the practice of inoculation caught on throughout the colonies.

Again, Boston's Puritans are at the core of a story—their openness to knowledge and respect for learning, coupled with their hidebound beliefs and censorial attitudes. This combination of social, political, and religious views—many of which seem contradictory—established a climate for the first use of inoculation, as did the rampant misery caused by the disease.

In the close-knit colony of seventeenth- and eighteenth-century Boston, smallpox was an everyday horror, almost comparable to the terror and suffering associated with plague. It surrounded the settlers. They could see the pockmarks, like gouges to the flesh, on the faces of those who had survived it. They could track the mounting numbers of epidemics (1702, 1721, 1730, 1752) by following the movements of clergymen, hurrying from house to house in their black cloaks, during their visits of consolation, unknowingly spreading the disease. Smallpox invaded politics, changing policies of settlement and immigration. The disease, its victims, and methods of control were not only reported upon in newspapers, but helped to differentiate newspapers. Like views on war in our own time, opinions on the advisability of inoculation were part of what distinguished newspapers from each other.

The disease invaded associations, dividing clergymen, physicians, and governors whose views of morality, community, divinity, and public health could not be separated from the practice of inoculation. As with issues of abortion and stem cell research, what some considered a medical matter was for others anathema.

Cotton Mather and his father, Increase, are caricatured in many histories of early New England. They represent the stern, hardline, inflexible Puritans. Cotton Mather is described in correspondence—in a social note, recommending that a gentleman make his acquaintance, no less—as "an excellent hater." Father and son were opposed to the popular celebration of Christmas (see page 195)—though Cotton accepted a purely religious observance of the holiday—and were virulent, sermonizing witch hunters, responsible for stirring up hysteria that brought about the harrowing deaths of numerous innocent women. But Cotton Mather, a graduate of Harvard at age fifteen, had considered studying medicine before deciding on the ministry, and he kept his hand in, reading medical journals and following the accomplishments and careers of the medical men of the day. Like a good doctor, and a good minister, he was curious and observant. In 1714, he had read a copy of the *Philosophical Transactions of the Royal Society of London*, which reported on the use of inoculation in Constantinople. In 1716, he read another report in the same journal and wrote to Dr. Woodward of the Royal Society of a conversation he'd had with his slave, Onesimus, who had told him of the widespread practice of inoculation in Africa. (Had these conversations between owner and slave been routine, the settlers would likely have known of the African practice almost a century before.)

Mather wrote to Dr. John Woodward:

I do assure you, that many months before I mett with any Intimations of treating ye *Small-Pox*, with ye Method of Inoculation, any where in *Europe*; I had from a Servant of my own, an Account of its being practised in *Africa*. Enquiring of my Negro-man *Onesimus*, who is a pretty intelligent Fellow, Whether he ever had ye *Small-Pox*; he answered, both, *Yes*, and *No*; and then told me, that he had undergone an *Operation*, which had given him something of ye *Small-Pox*, & would forever Praeserve him from it; adding, That it was often used among ye *Guramantese* [group of people], & whoever had ye Courage to use it, was forever free from ye fear of the Contagion.

In *Princes and Peasants: Smallpox in History*, author Donald R. Hopkins describes the campaign Cotton Mather then initiated to convince his community to adopt the practice of inoculation. (The quotations in this essay are all from Hopkins's book.) There were but ten physicians in Boston at the time, all wary of Mather, who was known to be difficult, obstinate, and—to use a modern expression—a man "with an attitude." But his thinking was rigorous, and his writing graceful and persuasive. With flattering correspondence and astute medical references, Mather convinced Dr. Boylston to try the first inoculation.

On June 26, 1721, Boylston took pus from a patient with smallpox, and using a "sharp toothpick and quill," inoculated his precious only son—Thomas, age six—and two black slaves. Thomas and the slaves developed mild infections. They lived. They became immune to smallpox.

Boston, the community Mather and Boylston hoped to serve, did not respond with elation, reverence, and awe, but with outrage and violence—against Boylston and Mather, attacking them in pamphlets, in the press, and in person. Colonists broke into Boylston's home, seeking to beat him, even to hang him. A homemade grenade was thrown into Mather's home, and he was the powerful minister of Boston's North Church!

For those opposed to inoculation in the community at large, the issue was the sacred obligation to community that Mather's promulgation of inoculation seemed to violate.

For the governing and professional elite, the issue was more one of politics, of whether the Puritan ministers "should continue to dominate virtually all aspects of the community's secular and religious life," as author Donald Hopkins puts it. Ironically, in this instance of public health, the position of the obstinate, dogmatic Puritan minister was the more liberal and enlightened.

Practically, the question was, did inoculation work? Did it keep people from getting deathly sick—or did its use cause the disease to spread?

In the midst of this public uproar and humiliation, Mather suffered a private, related agony. His son's roommate at Harvard had smallpox. The two boys had breathed the same air, shared the same table and the same books for months. In the heat of the controversy, Samuel Mather asked to be inoculated. On August 1, 1721, Cotton Mather wrote in his diary:

> Full of Distress about *Sammy*; He begs to have his life saved, by receiving the *Small-pox*, in the way of *Inoculation*... and if he should after all dy by receiving it in the common Way, how can I answer it? On the other Side, our People, who have Satan remarkably filling their Hearts and their Tongues, will go on with infinite Prejudices against me and my Ministry, if I suffer this Operation upon the child.

Young Sammy was secretly inoculated. He became mildly ill, recovered, and gained immunity. He was the only one of Cotton Mather's sixteen children to outlive him.

The controversy continued, as did smallpox. But in time it became apparent, and incontestable, that while some people died of inoculation, most did not, and they became immune. The risk of dying from inoculation was much lower than from contracting smallpox.

By the spring of 1722, Dr. Boylston had inoculated almost 250 citizens—in defiance of an order of Boston's selectmen. Only 2.4 percent of his patients died, according to Hopkins in *Princes and Peasants*. In 1726 he was elected a member of the Royal Society and published *An Historical Account of the Small-pox Inoculated in New England*. He practiced medicine until he was over seventy years old. Cotton Mather became the first native-born American to become a fellow of the Royal Society.

Inoculation was introduced to Philadelphia in 1730, New York in 1731, and Charlestown in 1738. By the end of the eighteenth century, it was widely practiced in Europe and America. Some 2 to

3 percent of those inoculated died, but the practice dramatically decreased fatalities. The English physician Edward Jenner (1749–1823) built on the practice of inoculation to develop a smallpox vaccination, which further reduced the incidence of the disease. Dr. Jenner's accomplishments were unknown to Cotton Mather, who died in 1728.

It is something to think about, that a combination of compassion, knowledge, and a wish to relieve suffering overcame prejudice and fear—in the mind and soul of one stubborn, supposedly hateful man, and later in many others. In no small way, inoculation was permitted because of that one man's persistent, eloquent, and adroit communication—the writing of letters in the face of bombs hurled into his room. Words halted disease.

ETHER

I n the hushed museum gallery, scores of visitors stand before an enormous oil painting depicting a medical miracle. The first surgery performed on a patient under the anesthesia of ether happened 150 years ago, but artist Thomas Eakins's 1876 depiction of the subject is as astonishing, vivid, and miraculous seeming as when the gory but glorious public experiment was first conducted.

Eakins's masterpiece *The Gross Clinic* was painted for the centennial of the United States, and was on view in 2002 at a retrospective of the artist's work at New York's Metropolitan Museum of Art. For its three-month run, from early morning until the museum closed its doors, this particular painting attracted a reverential group of viewers. It depicts Dr. Samuel David Gross, a pioneering surgeon, operating on an anesthetized patient at Jefferson Medical College in Philadelphia. The operating room is dark; shafts of light shine on Dr. Gross's face, his bloody hands, and the hands of the other doctors, all intent on caring for their patient. Silent before the masterpiece, some museum visitors wept. Others caught their breath—at the artistic command of the painting, at the scientific event depicted, and at the amazing, almost mystical, phenomenon that a human being or an animal can, with anesthesia, be sliced open with knives—organs lifted and shifted, sutured, repaired, and removed—without ever feeling pain.

A similar response—of wonder, awe, muted terror, gratitude—greeted the first public surgery performed under the anesthesia of ether at Massachusetts General Hospital (MGH) on October 16, 1846, thirty years before Eakins did his painting commemorating a similar surgery in Philadelphia.

Ether was not invented in Boston, nor used here for the first time on a patient. It was discovered by a Spanish chemist in 1275, synthesized by a German chemist in 1540, and soon after, its hypnotic effects noted by a Swiss scientist. By the early nineteenth century, American doctors were using the clear liquid to treat pulmonary disorders. It was also a recreational drug, used at "ether parties" where nineteenth-century medical students generally first made its acquaintance.

The dramatic surgery of 1846 conducted in the Bulfinch building—the hospital's early home, and today known as the Ether Dome—marked the first public demonstration of ether used for anesthesia during a surgical procedure. Never before had such an experiment been successfully conducted in such a prestigious public setting, closely monitored by leading physicians.

At 10:15 a.m., Dr. William Morton, a twenty-seven-year-old Massachusetts dentist who had been using ether on his patients, administered vaporized ether to Edward Gilbert Abbott. Abbot was a twenty-year-old printer with a congenital vascular malformation, a tangling of arteries and veins. Dr. John Collins Warren, the eminent senior surgeon at Massachusetts General Hospital—skeptical of Dr. Morton's ether and waiting for his patient to jump, or scream—lifted his scalpel and incised Abbott's neck. The patient did not move. He appeared to be asleep. "Gentleman, this is no humbug," said Dr. Warren, scalpel poised. He removed Abbot's problem—the tangled blood vessels—and sewed him up. Later, the recovered patient—speaking crisply as though for the evening news—remarked, "I did not experience pain at any time, though I knew the operation was proceeding." Anesthesia was launched.

Though the trajectory of the discovery is unclear, the format and forum in which ether was successfully demonstrated and later promulgated at Massachusetts General Hospital are typically Bostonian. By the mid-nineteenth century, the city was a center of knowledge, including science and medicine, with a well-established professional network. The prevailing sensibility lent itself to careful, rational assessment of phenomena and application of theoretical knowledge. As a citadel of medical academia, Boston was a crossroads for information exchange, networking, mentoring, and getting one's chops. If you could make it here, you could make it anywhere.

Massachusetts General Hospital became a world center for ether anesthesia and continued to develop anesthesia Firsts. During the 1920s, the "anesthetic record" was introduced. Prior to this, doctors and nurses had not recorded changes in patients' blood pressure, respiration, and pulse under anesthesia. During the 1940s, MGH first applied anesthesia techniques to supporting severely injured trauma and burn patients. In the 1950s, methods were developed to assess and track patients' pain levels. For decades MGH has led in developing and testing new neuromuscular blocking agents and investigating the molecular biology of the nervous system's response to pain.

But oddly and tragically, like some Gothic horror tale, the invention that was to be of incalculable benefit to so many patients was associated with tragedy for several scientists associated with its invention. A public feud developed among the four men who claimed to be responsible for the discovery, especially between Dr. Morton and Dr. Charles Jackson, a chemist, physician, and professor of medicine, who had prepared Dr. Morton, a dentist, for medical school admission. After twenty years of struggle to patent ether, Dr. Morton, who had claimed full credit, died in poverty at age forty-nine, either of stroke or heatstroke, leaving his neglected family in ruin. Dr. Jackson was felled by a paralyzing stroke, diagnosed as insane—which he probably was not—then committed to McLean Asylum. Dr. Horace Wells—a Connecticut dentist, Dr. Morton's former partner—became addicted to chloroform, threw acid at a prostitute, was jailed, and committed suicide in prison at age thirty-three. Dr. Crawford Washington Long, a Georgia physician who had successfully used ether on a patient back in 1842, died at age sixty-two in the midst of administering ether to a farmwife in labor.

This glory and controversy are reflected in a massive granite and red marble monument in Boston's Public Garden. Erected in 1867, it stands near the corner of Arlington and Beacon Streets, the only statue dedicated to a pharmaceutical discovery. The monument is suitably Gothic in appearance. Its inscription, by omission, testifies to the uncertain credit for the first public use of ether in surgery. In the Bostonian fashion of diplomatic understatement, sculptor John Quincy Adams Ward dealt with the issue by not naming any discoverer. Instead, he carved the Good Samaritan. The irrepressible Dr. Oliver Wendell Holmes, whose oratory was not limited to dedications of the Boston Public Library (see page 117), dubbed it a memorial "to ether—or either."

Still, the looming monument commands attention to the stark mercy of anesthesia. A relief showing an angel of mercy, descending to earth, bears this inscription from the Book of Revelation: "Neither shall there be any more pain."

SUBWAY

❦ 1 8 9 7 ❧

Bostonians, New Yorkers, Londoners, Parisians—urbanites all —can't imagine life without the subway: descending into a tunnel, traveling underground, surfacing soon after, and—*voila!*— arriving at one's destination. Yet these underground roadways for the carless, formerly coachless, are only about a hundred years old. Boston's subway, opened in 1897, was the first in the U.S. The project took decades to shape, but once the first spade struck Boston Common, it took just two years to construct. It made a huge mess in the heart of the city, but ease and convenience descended when Park Street station opened.

The city had been suffocated by streetcars. Boston's streets are crooked and narrow, and every other block seems to hold a precious something that must not be tampered with. By the late nineteenth century, this city of precious somethings was in gridlock. There were plenty of streetcars, but they were pulled by horses; it sometimes took over half an hour to travel one mile.

The city, citizens, and businesses searched for solutions. A great pageant of politics and public works proposals played out from the late 1870s, when proposals for elevated lines were first considered, to the late 1890s, when compromise legislation (for both elevated lines and subways) went to a referendum. The people of Boston voted, the modernists won—by a narrow margin—and construction began. Two years of hard labor began; men and boys digging the tunnel by hand, steam-powered vehicles moving the landfill through the downtown. A horrific gas explosion at the corner of Boylston and Tremont Streets killed nine workers, injured many more, and terrified those who lived and worked in the zone.

The pageant played out much the way such projects do in Boston today. There was a problem. There was no obvious applicable solution. The situation was allowed to reach near catastrophic dimensions until the people, as opposed to the government, demanded a solution. A few bright minds—engineers, innovators, entrepreneurs—stepped forward. Proposals were offered, debated, rejected. Opinions divided along class lines. The problem worsened, the citizens clamored, the naysayers (real estate values will plummet) and scaredy-cats (we will be poisoned by noxious, underground vapors) were mollified. The legislature got to work. Commissions were established. Science and business married. The public voted yes. Contracts were signed, money changed hands, work began! The people who voted against construction complained. The people who voted for construction complained.

Newspapers followed every step. At long last, page one of the September 1, 1897, morning edition of the *Boston Globe* ran an all-caps headline: CARS NOW RUNNING IN THE SUBWAY. The story was preceded by five elegant drawings of subway entrances, station interiors, and a ventilator (looking like a vestige of Stonehenge) on the Common, nicely graced by a shade tree. The news story began:

> The Boston subway is open for public travel; underground transit is no longer a dream in the hub.
>
> Before the readers of The Globe, or at least the majority of them, open their morning paper today, the hum of the electrics will have turned the underground railway into a scene of human activity wholly new to the busy life of Boston.

The article detailed the cosmopolitan populace "whirled along at a rate of speed which they have never experienced in the same locality before, and projected out of the earth at the Park Street exit of the subway."

The *New York Times* editorially sniffed (just before the subway opened), "That so conservative an American town should happen to be the pioneer in adopting this is viewed as remarkable.... Boston, after the way of a venerable dame who survives the shock of moving from the ancestral cottage to a modern residence, began to look upon the impudent innovation with a degree of complacence."

The "impudent innovation" did not come easily to Boston. As traffic worsened, several solutions had been entertained, including the idea of an elevated railway, examples of which already existed in New York, Chicago, and Brooklyn. Subways were not unknown: London's underground had existed since 1863, Glasgow's since 1886, and Budapest's since 1896.

In 1887, the Massachusetts state legislature consolidated several streetcar companies into the West End Street Railway. In 1888, the city of Richmond, Virginia, converted all its horse-drawn streetcars to electrically operated vehicles. Henry H. Whitney, a magnate of horse-drawn streetcars and head of Boston's West End Street Railway, made several trips to Richmond, was pleased by what he saw, and in 1887 electrified most of his enterprise. Soon more streetcars clogged Boston streets, and now they gave off sparks!

In 1891, the legislature appointed a rapid transit commission to study the problem. In 1893, yet another commission advised both a network of elevated railways and a subway tunnel, to get the mess of trolleys off the street. The recommendation created an uproar. The logical place for a tunnel was the area of greatest traffic, the Boston Common. Many members of the oldest, wealthiest families were horrified by the notion of disturbing this hallowed ground. (Indeed, over the years, construction workers would "accidentally exhume" the remains of almost one thousand colonial settlers. Their bodies were moved to the existing Central Burying Ground in Boston Common.) The beauteous Public Garden, which had finally come into its own, would incur the loss of nicely established trees. The interruption of commerce—shades of the Big Dig—was also an issue. It was estimated that construction would take eighteen to forty-two months, and traffic would be stopped at Boylston and Tremont Street, the hub of the hub. Businesses would fail, workers would be unable to reach their jobs. Property owners were convinced that their buildings would implode. There was also a fear of being underground during transport—waiting for trains, walking about, and breathing in a region of sewers, gas mains, snakes, worms, mice, rats.

Following Bostonians' yes vote on the critical referendum in 1894, the project moved ahead. Construction began in 1895. A tunnel was dug. Beneath Boston Common, it ran along Tremont Street from Park Street to Boylston Street, with entrances at Park Street (as today) and on the Boylston Street side of the Public Garden.

Though by modern standards the first subway tunnel was short, disruption in the core of the city during these last years of the nineteenth century felt to Bostonians of the day much as the Big Dig did to Bostonians at the end of the twentieth century. In the heart of the city, huge amounts of earth were moved from under the Common; small steam-driven trains operating on an overhead trestle moved the fill. The hub of the hub was interrupted for two years: noise, dirt, disruption, massive inconvenience, rerouting, and loss of time. Negotiation between the city and the public ensued. The modern practice of mitigation—agreements between municipalities and the private sectors to compensate for damages —took shape. Construction was restricted during workday hours, between 7:00 a.m. and 7:00 p.m. (two-thirds of the street and all surface car tracks were kept open). Even at night, when most construction would be done, one-third of the street remained open.

The transit commission also agreed to restore the Public Garden, according to a plan by premier landscape architects Olmsted, Olmsted & Eliot.

By the turn of the century, the subway was a fact of life. Bostonians descended, rode, ascended, and arrived, mainly without incident. "Previously, 200 trolleys often plied Tremont Street in each direction every hour, to hopeless confusion and the ruination of schedules," notes Brian J. Cudahy in *Change at Park Street Under: The Story of Boston's Subways*. "A month after the subway opened, it comfortably took care of 282 cars per hour." Extensions were added to the line—ingenious combinations of elevated and subway lines to Roxbury, East Boston, Charlestown, and Cambridge. The fare was five cents for two decades, bumped up to eight cents in 1918.

In the decades that followed, it all seemed to function well, this special-for-Boston's-streets-and-sensibility combination of old and new: roads, tunnels, elevated lines, electric, steam, horses, carts, and wagons. But every so often something bizarre and altogether Bostonian would happen, a collision of worlds.

On a cold December night in 1917, Daniel Kinnally of Chelsea, a deliveryman, was finishing his long workday. He left his horse and wagon unhitched—it was late and he needed only to make a quick drop in an apartment building. The horse, who had also had a long, cold workday, soon took off with his wagon in tow, cantering to Kenmore Square. Along the pitched roadway into the underground he went, clacking along the tracks. The authorities alerted the transit agent at Massachusetts Avenue, the next stop. Soon, out of the darkness galloped the terrified horse and the racket of his cart, "looking for all the world like a fugitive from Ben Hur," as Brian Cudahy tells it. As the transit employees still lived in a world where horses were familiar creatures, agent David Berry was able to flag down the animal, escort him above ground, and return him to deliveryman Daniel Kinnally. To add to the surreality of the evening, the shivering deliveryman had been wandering about for hours searching for his horse in the vicinity of a seeming apparition, a Venetian palace, rising in the gloom. That would be Fenway Court, the home of Isabella Stewart Gardner, completed just fifteen years before. The building of the subway (1895–1897) and Fenway Court (1900–1902) almost overlapped, the city spiraling into modernity, and antiquity.

KIDNEY TRANSPLANT

❦ 1 9 5 4 ❦

In 1954, Richard Herrick, a small-town boy who grew up in Rutland, Massachusetts, lay dying in the U.S. Public Health Service Hospital in Brighton. The once vigorous twenty-three-year-old Coast Guard veteran knew what he had—incurable kidney disease —and that he would soon die. His grandfather was a doctor and had explained the sorrowful situation; there was no cure for nephritis. One day, Richard's twin brother, Ronald, was visiting, along with his older brother, Van, and younger sister, Virginia. Richard was having a particularly miserable day, and Van was feeling more heartsick and exasperated than usual. He approached Richard's physician, Dr. David C. Miller, and asked if he couldn't just give his brother one of his own kidneys. Well, no, said Dr. Miller; it doesn't work that way. You can't just cut out an organ and splice it into another person's body. The recipient rejects alien tissue.

All the siblings were close. Dr. Miller couldn't stop thinking about them and the older brother's haunting question. He remembered that Richard was a twin, and that transplant experiments had been done at Boston's Peter Bent Brigham Hospital, mainly on dogs. He went straightaway to the hospital and told Dr. Joseph E. Murray about the dying young man and his twin brother. Dr. Murray had been working on the idea of transplant surgery for two years and was of the opinion that a transplanted kidney could function if immune system rejection problems could be solved.

That Richard and Ronald were identical twins could make all the difference. A patient's immune system might not reject the organ of his identical twin.

Ronald Herrick agreed to donate a kidney. A team of Boston physicians was assembled at the Peter Bent Brigham (today Brigham and Women's Hospital): Drs. John P. Merrill (head of nephrology), J. Hartwell Harrison (chief of urology), and Gustave Dammin (pathologist-in-chief). Dr. Henry M. Fox, chief of psychiatry, was one of several physicians also monitoring the case. The team paid close attention to moral and ethical issues, conferring with clergymen of different religions, seeking their views on the ethics of organ transplantation.

On December 20, 1954, Dr. Murray and his wife, Bobby, were hosting a large Christmas party at their home. Dr. Murray, lining up the eggnog ingredients and instruments, received a call that a cadaver he had requested was available. Murray turned away from the eggnog and festive Christmas bowl, drove through the icy streets to the post-mortem room, and went through the entire kidney transplant operation—a "test run"—with Dr. Francis Moore. He returned home for eggnog. Three days later, Murray, the lead surgeon, operating with Dr. John Merrill, took the kidney just removed from Ronald Herrick—it had traveled unceremoniously down the hospital hallway swathed in a wet towel in a bowl—and placed it into the abdominal cavity of Richard Herrick. If successful, Richard's iliac artery would bring fresh blood into its new kidney, and his vein would return used blood to the transplanted kidney. In his book, *Surgery of the Soul: Reflections on a Curious Career*, published in 2001, Dr. Murray, in his eighties, remembers the quiet reckoning:

> There was a collective hush in the operating room as we gently removed the clamps from the vessels newly attached to the donor kidney. As blood flow was restored, Richard's new kidney began to become engorged and turn pink. The

donor kidney had been without blood flow for a total of 1 hour and 22 minutes. There were grins all around. We removed the remaining clamp from the common iliac artery approximately 10 minutes later and immediately noted pulsation in Richard's right foot.

The kidney lay comfortably in its new site, pulsing with blood and showing pinpoint areas of bleeding on the surface. Urine flowed so briskly from the ureteral catheter that it had to be mopped up from the floor. Judging by that measure, the kidney was working perfectly.

This was the first successful kidney transplant—the first successful human organ transplant, period. It not only returned Richard to health, and did no harm to his donor-brother (who was alive and well, living in Maine, on the occasion of the fiftieth anniversary of the transplant in December 2004), but paved the way for hundreds of thousands of later transplants. In the five decades that followed, transplants would be made not only of the kidney, but of the heart, lung, and pancreas. Improved methods of dialysis followed, kidney banks were developed, and research was devoted to finding drugs that would suppress immune system reaction. It would become commonplace for laypeople to participate in decisions regarding their own medical care.

As with other medical advances, a confluence of circumstances and events led to the accomplishment in Boston. It had been a long time coming; Boston medical men and women had been involved in research into kidney disease for decades. Their consortium had shared information, even across continents.

In 1990, Dr. Joseph Murray, professor emeritus of plastic surgery at the Brigham, received the Nobel Prize in medicine for his pioneering work, founded in curiosity and a pledge to heal. As further testimony to Boston as a medical citadel, Murray shared the $700,000 prize with Dr. E. Donnall Thomas, who in 1956 had been the first to perform a successful bone marrow transplant. Both men had graduated from Harvard Medical School and were fellow residents at the Brigham.

The transplant story had almost everything going for it. In addition to the medical advance, there were numerous human interest elements and a raft of ethical issues. The two twins were best friends. They grew up on the family farm in rural Massachusetts and joined the military during the Korean War—Ronald the army, Richard the Coast Guard. Their mother, Marjorie Helen Herrick, died while they were in high school, and their father, Van Buren Herrick, while they were in the military. They had planned to move in with their aunt and uncle in Marlboro following their military service. But Richard was detained, his kidney disease diagnosed by a military doctor. He was moved from a Marine hospital in Chicago to Brighton, to be in the company of his siblings as he lay dying.

The hospital and transplant team had tried to keep the pending surgery confidential. But for all their medical knowledge, they were ignorant of the custom of reporters loitering at police stations waiting for a story, particularly a grisly murder or a celebrity picked up for a lurid misdeed. Dr. Murray had taken the unusual step of having the Herrick twins fingerprinted to ensure they were identical twins; he sent them to the local police station. The police beat reporters—possibly smoking cigarettes and wearing porkpie hats—pounced:

"Twin's Life May Hang on Fingerprint Today" read the resulting headline in the *Boston Herald.* In perfect journalistic fashion, the story was encapsulated in paragraph one of the nine-paragraph article: "A fingerprinting at 4 p.m. today at Roxbury Crossing police station may save the life of a 23-year-old Marlboro man seriously ill in Peter Bent Brigham Hospital." Murray and his team were thereafter under close public scrutiny, the aroused press hovering over their white-coated shoulders.

Dr. Murray himself was a focus. A deeply kind man, spiritual and observant, he painstakingly examined the ethical issues associated with the transplant. These dilemmas derived from the medical imperative to "do no harm." At issue was the efficacy of causing risk to a healthy patient by removing an organ. It was necessary to balance this risk with the potential aid to a dangerously ill patient, and to ensure that participants would receive as much counseling and information on their decision as possible. "We wanted everybody to know we were not doing anything frivolous or thoughtless," said Murray, in an Associated Press article in December 2004. Some people considered organ removal a desecration of the body and had criticized the doctors for "playing God."

Post-op, the ever loitering pack of reporters tracked yet another story—a romance. Following the successful surgery, Richard Herrick, an attractive young man on the mend, and needing weeks of care, fell in love with his nurse, Claire Burta, who supervised the ten-bed recovery room. A Canadian girl (grown women were still called girls in 1954), Claire did not return home to Nova Scotia that Christmas, and volunteered to watch over her patients. When Richard Herrick left the hospital and had sufficiently recovered, he called on his former nurse. The couple married and had two daughters.

Richard Herrick died eight years after the surgery. New problems developed in the transplanted kidney. Clare Herrick never remarried. "I loved one guy in my life, and that was Richard," she told Associated Press reporter Glenn Adams in 2004. "I just cherished the memories."

In the fall of 2004, on the occasion of the fiftieth anniversary of the successful transplant, Dr. Murray was interviewed at a time when stem cell research was much in the news—in the Massachusetts state legislature and in the U.S. Congress. "The use of stem cells is an indication of mankind's innate sense of curiosity," said Murray, perhaps remembering accusations of "playing God" that had been directed at his team. "We shouldn't stifle it at all."

From Dr. Murray's leap of faith and his medical skills, from the bonds between two brothers, and those between doctor and patients, hundreds of thousands of life-saving surgeries have been performed since Dr. Murray's First, December 23, 1954.

ART &
ARCHITECTURE

BOSTON LIGHT

❦ 1 7 1 6 ❦

If you sail into Boston Harbor, or even chug along in a ferry, there is a magical moment when the present-day city slips away. You can still make out the Boston skyline—a ribbon of architecture, including the mirror-surfaced Hancock Tower and gleaming H-shaped Federal Reserve. But the expanse of rolling water is so great; the reefs, sky, and gulls so compelling; the movement of the boat so lulling and insistent, you sense yourself in another time. Head two miles out—into Massachusetts Bay near Little Brewster Island—and your time travel is complete. You will fancy yourself navigating into a natural harbor some three hundred years ago.

This is no illusion. The gleaming lighthouse at Little Brewster, radiant white on sunny days, was the first lighthouse built in the colonies. It looks precisely as we would like a lighthouse to look, the essence of a beacon, sturdy yet graceful with a powerful beam. Boston Light separates land from sea, man from fish, the dangers of the depths from the order of civil society. During the day, Boston Light on Little Brewster Island looks like a tiny fishing village, with its houses and outbuildings on the shoals of a small island. At night it turns majestic, a shaft of light beaming over Massachusetts Bay, illuminating the ocean and smoke-colored sky.

It was not erected due to routine civil ordinance or lofty orders from the Crown. Boston merchants lobbied hard for it, reasoning that a proper light would distinguish their harbor. Boston was not the only port in the colonies, not the only site that would benefit from a beacon. Lighthouses existed overseas. But the necessary chemistry—a combination of civility and concern, an understanding of commerce and investment, and respect for science and invention—led to the building of the first American lighthouse here.

It was lit September 14, 1716. What we see today is actually a replacement for what the British destroyed. The demise of the original lighthouse is a tad complex. After the Brits commandeered it during the Revolutionary War, a group of rebels raided the lighthouse, setting it afire. The British stationed marines and commenced repairs. A few weeks later, under orders from General Washington, a three-hundred-man party stormed the island and again fired on the lighthouse, making it largely unusable. The following March, withdrawing from Boston, the British set a timed charge in the tower, destroying Boston Light. Its demise was a moral blow to the patriots, and created a navigational hazard. In 1783, the Commonwealth of Massachusetts authorized construction of a new lighthouse, which followed the original plan and incorporated a wall of the old tower. The cherished, restored lighthouse was operated by the Commonwealth until 1790 and then ceded to the United States of America.

Lighthouses were critical in pre-radio, pre-radar, pre–global positioning days—always targets for improved technology—and so Boston Light figures prominently in a sea of other Firsts.

The brightness of its light was of paramount importance. In the beginning, whale oil was used to keep sixteen "spider lamps"— lamps with leglike extensions mounted on a circular frame—illuminated. Ventilation was terrible in the lantern room, and great amounts of noxious soot and moisture accumulated, blocking transmission of light and sickening the lighthouse keepers. In 1790, keeper Benjamin Lincoln redesigned the lamp and improved the circulation of air. In 1810–1811, Winslow Lewis installed the vastly superior Argand lamps, which he quite literally called his own. A dramatic civil ceremony was held—the old oil lamps of

Boston Light extinguished, the new Argands turned on—and Lewis basked in the limelight.

A well-respected mariner and businessman, member of the Boston Marine Society, Lewis shamelessly patented his invention for a "reflecting and magnifying lantern," an oil-burning system that used a hollow circular wick in a glass chimney. Impressive-sounding, indeed—but Lewis lied. The lantern he introduced had been in use in Great Britain for twenty-five years. In claiming originality for a slightly modified version of an existing technology, Lewis anticipated the profiteering of today's drug companies by almost three centuries, perhaps yet another Boston First.

Later innovations included the installation of gas lights and Fresnel lamps, named for the French physicist Augustin-Jean Fresnel (1788–1827), an optical system that used a thin bull's-eye lens surrounded by concentric prisms. In 1962, a 1,500-watt, electrically powered bulb was installed in the center of a twelve-sided Fresnel lens. The beam flashes once every ten seconds, throws two million candlepower of light, and is visible for sixteen miles.

Boston Light had the first foghorn, too, though not the type you might imagine. In 1719, the first keeper of the lighthouse requested a cannon from the Commonwealth. Request granted. For the next 132 years, the stalwart cannon, which still stands at the site, boomed in warning as fog descended—announcing reduced visibility to the mariner, and impending doom to landlubbers with far-flung imaginations. In 1851, a bell was substituted for the cannon, retiring the oldest fog signal in the nation. To augment the bell, the station would also employ an eclectic, somewhat unmusical orchestra—a pre–jazz era jam of trumpets, sirens, and horns.

Boston Light is another kind of monument, a tribute to the historic pride that leads to political activism. When the Coast Guard attempted to un-man Boston Light during the late 1980s, as part of a national plan, local opposition was so fierce that the Coast Guard had to retreat—by land, sea, and congressional appropriation. Propelled by preservationists such as the Friends of the Boston Harbor Islands, Senator Edward Kennedy sponsored a bill to keep Boston Light manned. He was successful.

Tragedy roils in the lighthouse's Firsts. In the beginning, a terrible accident befell George Worthylake, the first keeper, and his family. He had been on the job just two years when a small vessel he was riding in capsized, drowning Worthylake, his wife, Ann, and younger daughter, Ruth, along with a friend and the friend's slave, Shadwell. On shore stood Ann Worthylake, George's older daughter, who watched the drowning of her family. The three members of the Worthylake family were buried at Copp's Hill Burying Ground in Boston's North End, one marker above their common grave.

Following the family's drowning, a former sloop captain, Robert Saunders, took over; he drowned just a few days into the job.

The memories of these early deaths have never left the pretty little island. Some say a ghost roams. It could be one of the drowned keepers or one of their family members, or one of the scores of sailors who have drowned on the island's shoals.

In our own time, old-timers remember how through the years of World War II, Boston Light was shut down lest it provide navigational aid to invaders. Those accustomed to seeing it at night, as regular as the moon—during ordinary activities such as walking their dogs, taking in wash, or closing a window when nights turned breezy—were reminded of the perils of war, on the other side of an ocean they could no longer see.

Today, the eighty-nine-foot white tower on its rocky promontory, a national historic landmark, is the last manned lighthouse in the U.S. Except that it is not manned, but womanned—for the first time in its history—by Sally Snowman, a member of the Coast Guard Auxiliary. She took her post in September 2003.

Snowman—a teacher and author, as well as a member of the volunteer force—dresses in seventeenth-century garb and acts as an interpreter for visitors; the lighthouse site is part of Boston Harbor Islands National Park. Though Boston Light has been automated since 1998, Snowman—a vigorous, outgoing woman—does pretty much what her forebears did, and not for the sake of historical reenactment. She checks instruments, reads reports, fixes leaky windows, and stays semialert through the night, "sleeping with one ear open."

Sally Snowman is the first female lighthouse keeper of the first lighthouse erected in what was to become the United States of America—the first lighthouse in the Western Hemisphere.

TONTINE CRESCENT

Our romantic idea of old Boston is wrong. But who could blame us? The evidence of the nineteenth-century city is so prevalent and pleasing—the wide, leafy boulevard of Commonwealth Avenue, the gold-domed State House and red-brick row houses—that we forget this is not the appearance of an older Boston. That early community would look like a backwater to us, a slapdash affair of cleared forest with small wood-frame houses. Everywhere dust, dung, debris. Even by the end of the eighteenth century, there were no planned residential neighborhoods, or brick sidewalks and promenades where a lady might safely and decorously walk.

But there were ideas. A few civic leaders and visionaries imagined how it could be: How it could look; how it might feel to be a Bostonian walking about, or on horseback, or in a coach in a planned cosmopolitan city; what the impact of buildings, streets, and landscapes could be on the individual and society; how space could be shaped to affect human life. Chief among these visionaries was Charles Bulfinch (1763–1844), citizen, architect, developer, urbanite, advocate of community.

Charles Bulfinch became one of the signature architects of Boston, converting the eighteenth-century pastureland, vacant lots, and seedy wharves into the sheen, polish, and urbanity of the nineteenth-century city. He was a gentleman of refinement, erudition, and verve, which translated into neoclassical buildings of elegance and restraint.

Bulfinch is known for his design of the Massachusetts State House, that citadel of Beacon Hill for over two hundred years; St. Stephen's Church, with its octagonal cupola, on Hanover Street in the North End; and grand houses for leading citizens such as Harrison Gray Otis, Boston's third mayor, senator, and man-about-town.

Bulfinch designed Massachusetts General Hospital—today's Bulfinch Pavilion and Ether Dome, the latter an operating theater with seating for students beneath a skylit dome, site of the first public demonstration of the use of ether as an anesthetic in 1846 (see page 81). Across the Charles River, University Hall and Stoughton Hall in Harvard Yard still mark his passage in our midst.

Bulfinch grew up in a wealthy family, the son of a physician, and graduated from Harvard as a classics major. He had been alert to his environment and interested in architecture since boyhood, and had studied architectural design, mainly on his own. Upon graduation, he made a Grand Tour of Europe, gathering impressions and ideas. At first his architectural studies were avocational, a gentleman's pursuit. But the loss of his family's fortune turned him into a workingman—an architect for hire and a public servant, a rather modern combination. His offices ranged from chief of police to head of the Board of Selectmen. He was a gentle, genteel fellow, judicious and fair, but with a wild streak when it came to building.

Two and a half centuries ago, as now, architecture was speculative work, relying on selling ideas, landing properties, rounding up investors. Outstanding citizen though he was, and well connected, Bulfinch was hounded by debt most of his life. His most visionary buildings bankrupted him. Yet to the end, he continued to try to build the city he imagined.

On his Grand Tour in 1785–1787, young Bulfinch was attracted to the well-established Continental cities and towns, especially in England, and fell head over heels in love with London and

Bath, where he saw genteel row houses, including those built in a curve. Though he loved his hometown—he grew up on Beacon Hill, in Bowdoin Square—in Europe he realized the architectural backwardness of Boston, and he returned home determined to elevate his own milieu to a place of humane elegance.

Bulfinch's vivid memories of curved English row houses led to the creation of one of his most magnificent contributions. It is gone (but for design descendents), demolished in the melee of commercial enterprise that characterized mid-nineteenth-century Boston. But so powerful was the building's design, its grace and urbanity, that it continued to influence. If you know where to look and what you are looking for, you can see a bit of its curve even now.

The 480-foot-long Tontine Crescent, built in 1794, was an elegant half-ellipse of buildings—sixteen painted-brick row houses connected in a curve, the first architectural crescent built in the U.S., and one of the first planned housing complexes in the new nation. The complex may even have been the first example of urban planning in the U.S., period. This gracious housing, which immediately created its own neighborhood, was built in what was then considered the South End (what we today consider downtown)—on the southern side of Franklin Place, with the western edge intersecting Hawley Street, the eastern edge near Federal Street.

The politicking and dealmaking that Bulfinch engaged in to create the Tontine Crescent would have made him no stranger to the realm of twentieth-century real estate development. To finance the project, he organized a group of investors in what was known as a Tontine scheme, named for a clever Neapolitan, Lorenzo Tonti. In this arrangement, individual investors bought shares of a property. As they died, their portfolios passed to the surviving investors, until all were held by a single vigorous, triumphant, ancient shareholder—the winner of a kind of real estate roulette.

Bulfinch's original idea—two crescents facing each other, a slim oval park in the middle—was never fully realized. Had it been, an elliptical arrangement of thirty-two elegant homes on Franklin

Place, divided into two semi-ellipses, would have been on either side of a verdant, elliptical park, a place of cushy lawns and vase-shaped shade trees. But even the existence of half the ellipse—a single crescent, all that financing would allow—exemplified an idea and ideal of what urban American life could be.

These were elegant three-story houses, painted gray to resemble stone, with crisply contrasting white pilasters, and arranged in two groups of eight houses on either side of a neoclassical archway.

If it is possible to speak of "buzz" in eighteenth-century Boston, the Tontine Crescent generated plenty. Like the Ritz condominium building in Boston, or the Dakota apartment building in New York City, the Tontine residents were a Who's Who of eighteenth-century Boston. Bulfinch might have made a fortune if the financing he used had been more customary, and if business ventures generally had not been suffering because of failed commercial treaty negotiations with the British.

In addition to these stylish residences—a community of polite neighbors living in a curve—Bulfinch designed two communal public buildings as part of the complex: Holy Cross Roman Catholic church and the Boston Theater. This was Boston's first theater, the first public building in which plays were performed—a resounding recreational rebuff to the Puritan legislators and the city's staid history. Before, plays had been performed privately, in homes, but to stage a show in public was a dramatic break with the past, a thunderclap, to put it theatrically. The construction of a Roman Catholic church, Boston's first, was also a bold move, politically as well as from a planning perspective.

And so, over two hundred years ago in Boston, an American architect created a planned urban village, with buildings designed for residence and family life, for culture and entertainment, and for worship, including the idea of drawing in worshippers of different faiths from those who occupied the residences. The setting combined public, private, and communal space; it promoted tolerance and exchange, and it was beautiful.

It would make an inspirational tale to say that the Tontine Crescent was the crowning glory of Bulfinch's career. The facts intrude. Its creation ruined him financially and brought grief to him and his family.

In *Bulfinch's Boston*, author Harold Kirker spins a sad and moving story of Bulfinch's financial apocalypse. He lost everything on the Tontine Crescent, including the dwindled remains of his and his wife's money and that of several relatives. The resulting bankruptcy forced all remaining funds to pass to his creditors. "When the disaster was behind him," Kirker writes, "and the life of his family irreparably altered, the architect bitterly reproached himself: 'With what remorse have I looked back on these events, when blindly gratifying a taste for a favorite pursuit, I envolved for life myself and wife with our children.'"

But Bulfinch had not pursued his calling blindly and selfishly; in the main, he had planned and built as an act of public generosity, to create a dream of a city. He never recovered financially, but continued to design, plan, build, and serve.

The generosity of Charles Bulfinch—his passion for civic involvement and improvement—was unstoppable, even after he had lost everything. A large room with Palladian windows, designed for a private library, was just over the central arch of the Crescent. Bulfinch gave the room to the Boston Library Society, a union of book collectors, and to the Historical Society. He had designed this elevated, light-filled space as a reading room, repository, and dispenser of knowledge. That his own life had darkened did not deter him from fulfilling his promise.

He also donated a precious classical urn he had collected in Europe. He had intended it to stand in the grand ellipse of his project, but he placed it at the center of the realized park—albeit a partial ellipse—dedicating it to Benjamin Franklin, a son of Boston, who had gone to school at Boston Latin, then located on nearby School Street. When the buildings were demolished in 1858, the urn Bulfinch had selected and cherished—now further smoothed by time and weather, and decorated with moss—was moved to his own grave in Mount Auburn Cemetery in Cambridge, America's first garden cemetery.

In spite of the Tontine debacle, Bulfinch became an activist —convincing city fathers to enact an ordinance to regulate building height and style—and a planner, almost in the modern sense. He devised the plan for Broad Street, which converted the motley, ragtag seventeenth-century wharves to a handsome street of nineteenth-century warehouses and stores. He designed Central Wharf, where fifty-four slender Federal-style buildings went up. He loved the city, and architecture, but also needed income and held municipal positions for decades.

Though the Tontine Crescent is gone, Franklin Street still curves, evoking the memory of an elegant experiment. Arch Street —off Summer Street near Downtown Crossing—is so named because it once passed through *the* arch, the centerpiece of the Crescent. There are more tangible memories, too. The Sears Crescent near today's Boston City Hall is a graceful reminder, and the façade of the Kirstein Business Branch of the Boston Public Library—also downtown, built 1929–1930—replicates the entire central portion of the Tontine. Hundreds of brick row houses all over town descend from Bulfinch's vision, including close cousins in Roxbury Highlands and Brookline. In *Built in Boston: City and Suburb, 1800–2000*, Douglass Shand-Tucci cites "a kind of Queen Anne Tontine Crescent of fifteen attached brick and half-timbered town houses," built by J. Williams Beal on Elm Hill Avenue in Roxbury Highlands (originally called Harris Wood Crescent!), and a block of fifteen red-brick connected houses on Beacon Street in Brookline, built in 1907 by Murdock Boyle, a Dorchester designer. These lovely blocks, which still exist, are acts of architectural veneration.

All over the built city, our connected communities carry the spirit of the youthful Charles Bulfinch, who had come back from his travels so dazzled and hopeful.

He first saw row houses over two centuries ago, during his sojourns in London and Bath, and he kept them alive in his memory. Later, he revived and reworked these memories, creating a series of spare, elegant row houses in his own hometown. Today, these dwellings—which seem to have been here forever, but which were transported and transposed in the mind of an architect—are an integral, beloved part of Boston.

In the metaphysics of architecture, perhaps every house ever built and every house we ever live in is one long memory, reconstructed.

USS *CONSTITUTION*

❦ 1 7 9 7 ❦

The USS *Constitution* brings tears to the eyes of the unsentimental. Her appearance causes landlubbers to yearn for ocean voyages. Her antiquity and experiences—fighting wars and preserving peace—inspire even history haters to hit the books.

She was the first U.S. Navy ship to be built in Boston, here in the North End at Edmund Hartt's shipyard. Her prowess in battle led to Boston's preeminence as a shipbuilding town. She is still docked in the Charlestown Naval Yard, once a major shipyard, and today home to Boston National Historical Park.

She is the oldest commissioned warship still afloat in the world. During the War of 1812, she was the first ship to win a battle with the British Navy, a force considered invincible.

Launched in 1797, never defeated in battle, repeatedly rescued from abandonment—once by a rousing poem of Oliver Wendell Holmes, another time by a national children's campaign—she is a monument to Boston history, and not only in nautical and military terms. Famous Bostonians are associated with her. New England trees and artisans helped to build her. Our seamen and women eased her in and out of ports, and charted her courses hither and yon.

She is still a sight to see: a majestic wooden vessel almost 300 feet long from bowsprit to taffrail, and as tall as a twenty-story building, her main mast towering 189 feet. Regrettably, she can no longer tolerate the stress of sails. Were she rigged, she would carry thirty-seven sails. (They were hand-stitched at the Granary area of Boston Common; see page 233.) No wonder she had such amazing speed and maneuverability in youth, and well into middle age. During one of her nineteenth-century transfers, she beat the speed of the tug!

Over two centuries ago, merchant ships of the new American nation were being harassed by pirates. In 1794, President George Washington signed a bill to authorize construction of six frigates (sailing warships mounted with guns) in six different ports, including Boston.

Designers of the USS *Constitution*—naval architects Joshua Humphreys and Josiah Fox—created a larger, faster ship than other frigates of the era. Her hull was remarkably sturdy, yet she was maneuverable and fleet. She carried a crew of 450 and was mounted with fifty-four pieces of artillery.

Her timbers are white oak from Massachusetts, New Hampshire, and New Jersey; yellow pine from Georgia and the Carolinas; and the remarkably durable live oak from the sea islands of Georgia. Her glorious masts are white pine from Unity, Maine. "Old Ironsides," she is called, though made of wood.

She earned her sobriquet in action. During the War of 1812, under the command of Captain Isaac Hull, she encountered the HMS *Guerriere* on Georges Bank, seven hundred miles off the Boston coast. She trashed and thrashed the great British ship, turning it into "a perfect wreck," according to contemporary accounts. A British seaman watched cannonballs battering the hull of the American ship and doing almost no damage. "Huzzah!" he cried, in astonishment. "Her sides are made of iron! See where the shot fell out of her!" She became "Old Ironsides" after that.

Old Ironsides' exploits are 1940s movie material. In 1803 she subdued the Barbary pirates off Africa's north coast. During the War of 1812, under the command of Captain William Bainbridge, she captured the frigate HMS *Java*, manned by a crew of over four hundred and mounted with forty-nine guns. In two hours of horrific fighting, she totaled the British vessel, shooting out every mast and spar.

In a genre of mannered, old-fashioned British plays, the same guests appear at seemingly unrelated garden parties. Similarly, the same characters—Paul Revere and Oliver Wendell Holmes, for example—appear repeatedly in Boston history. The copper-sheath bottom, bolts, and spikes of Old Ironsides were made by Paul Revere. And when early danger threatened her status as a vessel to be treasured, the literary skills, love of Boston, and talent for publicity of Oliver Wendell Holmes rescued her. She was constructed of wood, and her life was limited. In 1830, the Navy was evaluating her future, and a newspaper article prematurely reported the ship as slated for destruction. Holmes, a law student at Harvard at the time, wrote a poem that was published all over the U.S.

Old Ironsides
Ay, tear her tattered ensign down!
Long has it waved on high,
And many an eye has danced to see
That banner in the sky;
Beneath it rung the battle shout,
And burst the cannon's roar;—
The meteor of the ocean air
Shall sweep the clouds no more.
Her decks, once red with heroes' blood,
Where knelt the vanquished foe,
When winds were hurrying o'er the flood,
And waves were white below,
No more shall feel the victor's tread,
Or know the conquered knee;
The harpies of the shore shall pluck
The eagle of the sea!
Oh, better that her shattered hulk
Should sink beneath the wave!

Her thunders shook the mighty deep,
And there should be her grave;
Nail to the mast her holy flag,
Set every threadbare sail,
And give her to the god of storms,
The lightning, and the gale!

A great hue and cry went up. Congress appropriated money to save her. Following repairs, she was used mainly symbolically, like an old admiral turned to diplomatic missions. She made many a stately appearance. In 1878–1879, she carried the American exhibits for the Paris World Exposition, her last cruise abroad.

In 1905, when the Navy decided to use her for target practice, Holmes's poem was again widely published (by then Old Ironsides was over one hundred years old, and Holmes's poem seventy-five!), and Congress appropriated $100,000. In 1925, as she became further degraded, schoolchildren across the U.S rallied, donating pennies to fund a restoration. During the later 1920s, she was extensively restored. Like a movie star, she then embarked on a grand tour, calling at ninety-one ports during the early 1930s, traveling as far south as Panama, and welcoming over 4.6 million visitors.

One last peril awaited. In 1974, during President Richard Nixon's administration—and following Massachusetts's lack of electoral support during the presidential election of 1972—the Charlestown Navy Yard was closed. But the 130-acre surplus Navy Yard eventually became condos, offices, stores, and a majestic national park, where the "eagle of the sea" is now forever protected and secure. You can visit her all year round, free of charge, in her home of two centuries, the Charlestown Navy Yard.

Huzzah!

BOSTON PUBLIC LIBRARY

❦ 1 8 5 4 , 1 8 5 8 , 1 8 9 5 ❦

Even those of us who barely recognize a brick, a beam, or a buttress are familiar with the "form follows function" injunction. We even understand what it means. An electric power plant is webbed with wires, a sewage treatment plant has huge unhidden ducts, a bank building is sturdy, impregnable, secure.

But what happens when "function" is an idea, such as the proffering of knowledge, beauty, and wisdom to the people of a city, especially the city's poor? What does the idea of knowledge, beauty, and wisdom—and its graceful facilitation—look like?

The majestic but welcoming edifice of the Boston Public Library in Copley Square looks like what it is and was meant to be: the idea that a city can offer its citizens a free public library with books they are welcome to read in a grand room, a building with murals and paintings they are invited to appreciate and enjoy, books for borrowing and taking home, books for reading in the parlor or at the kitchen table at the end of a long workday—a day that might otherwise have lacked all edification, stimulation, and allure.

The idea of such a library—"a palace for the people," it has been called—was radical.

In mid-nineteenth-century Boston, collections of books belonged to wealthy individuals, colleges, religious institutions, and prestigious membership organizations such as the Boston Athenaeum. But for the hoi polloi, there was nothing. The exception might be a kindly bookseller willing to lend a bright young lad a book to read in a drafty back room at night. (Stay tuned, this is not merely a proverbial tale.)

Like other great ideas that germinated in Boston, the idea of a free, centralized, municipally funded library took a while to root, to find the right combination of players and promulgators, circumstances and cash. It is often thus in Boston, a kind of push-and-pull atmosphere, intellectual openness to new ideas impeded from realization by a respect for tradition, a resistance to discarding anything proven, useful, enduring.

It took over sixty years for the Boston Public Library as we know it to go from idea to Copley Square edifice with lanterns and lions couchant. In 1826, George Ticknor, the Smith Professor of the French and Spanish Languages in Harvard College and trustee of the Boston Athenaeum, suggested the essential idea in a letter to his friend Daniel Webster. Walter Muir Whitehill extracts portions of the letter in his *Boston Public Library: A Centennial History.* Ticknor's idea: to unify all the smaller private libraries in the city, allow books to circulate, and offer popular literature as a hook, a device to promulgate a love of reading. What followed were suitable murmurings about the democratic magnificence of Ticknor's concept. But the book barons and bureaucrats of individual libraries were uninterested in ceding their fiefdoms. Nothing much happened for fifteen years—until the French ventriloquist and the free-thinking mayor teamed up.

In 1841, Monsieur Alexandre Vattemare, a man possessed of a theatrical personality, not to mention a trunkful of guises and props, arrived in Boston as part of his American tour to promote the establishment of libraries and museums. While his presentations were considered *de trop* by some, he had a great supporter in Mayor Josiah Quincy, who was also president of the Boston Athenaeum. With some public support, the maestro and the mayor organized three donations of books from the City of Paris, a bastion of enlightenment that already had a public library. In 1848, the Parisian donations and an "anonymous" contribution of five thousand dollars (from Mayor Quincy) led to an "Authorization to Establish a Public Library" by the Massachusetts legislature. It was the first act of its kind in the nation.

Many local citizens began to contribute books, including the accomplished, energetic, highly credentialed Professor Edward Everett, Harvard master of arts, minister of the Brattle Street Church, professor of Greek literature at Harvard College, congressman, former governor of Massachusetts, and former president of Harvard. Everett, a great champion of the state's public school system, recommended that a sturdy but plain building be at once constructed to house readers and books.

In 1854, the city consigned space for a public library: two rooms in a schoolhouse on Mason Street. In this modest home, 35,000 books circulated in the first six months in a collection of just 16,000 volumes! By the end of 1854, planning began for construction of a true library, designated for the purpose, on Boylston Street across from Boston Common. The new building, a handsome Romantic-style edifice based on plans by Charles Kirk Kirby and erected in 1858, was—like its Mason Street schoolhouse predecessor—a runaway success. About a year after it opened, more than thirteen thousand readers had registered to use its holdings. Approximately 15,000 volumes had circulated about 180,000 times, the equivalent of each and every book being checked out every month!

The impact of a public library—intellectual power to the people —was immediate, and also stretched into the future, transforming the lives of individuals, families, and communities. It was, in its own way and day, the Internet—but the Internet for *free*, and eventually housed in a beautiful, inspiring setting with a courtyard and works of art, abundant peace and quiet, comfortable chairs, and perfect light, all dedicated to reading.

Decades later, what is now known as the McKim building (as distinct from the sleek Philip Johnson addition in 1971)—the Boston Public Library of our own era—arose in Copley Square. Designed by Charles Follen McKim of McKim, Mead & White, built between 1888 and 1895, the Boston Public Library—aka the BPL —integrated art and architecture in a people's palace. "FREE-TO-ALL" reads the inscription in raised, carved granite letters above the central arch. The Italian palace of books and art sparked a national craze for Renaissance Revival architecture, and created a hub, a plaza, a square, with its existing neighbors, Trinity Church (1877), New Old South Church (1875), and the original Museum of Fine Arts building (1876), razed to build today's Copley Plaza Hotel.

On the grand day when the cornerstone for the McKim building was put in place—in a way, sixty years in the carving—Dr. Oliver Wendell Holmes read an eleven-stanza poem he had created for the occasion. The eighth stanza reads:

> Behind the ever-open gate
> No pikes shall fence a crumbling throne,
> No lackeys cringe, no courtiers wait,
> This palace is the people's own!

A paean, an ode, or a book could be written about the BPL's early patrons, and a movie made about its humble users turned movers and shakers and scholars, their intellectual horizons shifted by the offerings of the library. One of the most compelling of these scenarios is the story of Joshua Bates, a London banker who had grown up poor but smart in Weymouth, Massachusetts. Bates was instrumental in moving the BPL from an idea to a reality. In 1852, he offered fifty thousand dollars, a small fortune, to buy books for the two-room incarnation of the first BPL. A fine tale is told—in a report prepared for the designation of the BPL as a City of Boston Landmark—of how Bates enclosed a private letter with his hefty donation, disclosing how as a young man toiling in Boston, he longed for books, but could neither afford to buy them nor to subscribe to a private library. A local bookstore, Hastings, Etheridge & Bliss, allowed him to spend his evenings reading within the shop.

In this letter, Bates writes, "it will not do to have the rooms in the proposed library much inferior to the rooms occupied for the object by the upper classes. Let the virtuous and industrious of the middle and mechanic class feel that there is not so much difference between them." In 1855, Bates—pleased with plans to construct a designated building—stepped forward again to offer twenty to thirty thousand dollars to buy books for the existing library.

Bates Hall, which many library patrons fondly call the Reading Room, is named for the library's generous, kindly, public-spirited benefactor. On the second floor of the McKim building, Bates Hall is a magnificent public chamber for reading, thinking, and writing. The great, hushed room, 50 feet high and 218 feet long, spans the entire front of the library and overlooks Copley Square.

The BPL combines Bates's vision (a grand reading room for all), with that of Ticknor (providing popular literature to cultivate a love of reading), with that of Everett (a public library should continue a public-school education). Architecturally, it is an American work of art, espousing Renaissance principles. It employs mainly American materials. The façade is faced in granite from Milford, Massachusetts. The window arches are of Tennessee marble, the rear of the building Perth Amboy brick. The Doric columns in the courtyard are white Tuckahoe, New York, marble with white Georgia marble for the spandrels, rosettes, cornice, and parapet. A man was sent to Cape Cod to search for the proper color of sand.

The building's artists created in traditional styles, but they, too, were mainly American: sculpture by Augustus St. Gaudens and Bela Pratt, the great bronze entrance doors by Daniel Chester French, the interior mural paintings by John Singer Sargent, and Edwin Abbey's *Quest for the Holy Grail* located—where else?—in the book request room. French painter Pierre Puvis de Chavannes painted murals in his studio in Paris, home of the Bibliothèque Sainte-Geneviève, primary inspiration for the BPL's designer, Charles Follen McKim.

In the years that followed the opening of the BPL's first incarnation on Boylston Street, many other Firsts followed, including the innovation of branch libraries. To extend the arm of the library into Boston's neighborhoods, satellites for readers were orbited in East Boston (1871), South Boston (1872), and Roxbury (1873). Branches continued to open; there are twenty-seven today.

On a midwinter morning, it is a fine thing and a fun thing to go into the East Boston branch on Meridian Street, to shake off the snow from one's parka, and observe a grand old painting of a clipper ship on the wall. At a table just below the painting, a Vietnamese student, grandson of so-called "boat people," is deep in thought on a day off from school. His book is an atlas of the world. He is reading, as Joshua Bates would note, "virtuously and industriously."

FREE-TO-ALL

AUTHOR'S NOTE

A decorous disagreement has long existed between the Boston Public Library and a few other libraries in New England that claim their own version of "first." These include the Peterborough Town Library of Peterborough, New Hampshire, which cites its institution as the oldest free library in the U.S., and the Franklin Public Library in Franklin, Massachusetts, which asserts that its institution is the first lending library in the U.S. Cases may be made for each of these claims. But the founding of small libraries in New England towns, regardless of how noble and praiseworthy, cannot compare with the magnitude of the founding of the Boston Public Library, which required the involvement of the state legislature, academia, and patrons in business and the arts, and the role the enterprise assumed in a city of millions, including workers, students, and immigrants.

ISABELLA STEWART GARDNER AND FENWAY COURT

ᔥ 1 9 0 3 ᔦ

S he is celebrated as the first American to build a collection of old masters, the first to own a Matisse, the first to own a Raphael, the first to build and curate a private, personal art museum—not to mention the one, the only, to build a Venetian palace in Boston. It could also be said that she is the first developer to risk large-scale construction in the Fenway, which at the turn of the century was newly filled land. And who else has worn Worth gowns with ropes of pearls dripping to her waist, one rope for every European holiday, gifts from her adoring husband?

But as with the conferring of all awards, these accolades do not suggest the real impact and import of Isabella Stewart Gardner (1840–1924). Delicate in appearance, a stevedore in stamina, she created a community for art and artists in Boston, and was a kind of one-woman anti-defamation league, associating easily and graciously with gays, Jews, blacks, foreigners, and members of all social classes. She was drawn by intelligence, insightfulness, erudition, charm, talent, poise, and wit, and very much by male beauty, in marble and in flesh. These were her inspiration for creating an international network and cultural society, and an unsurpassed collection. Her Venetian palace, Fenway Court, is today the Isabella Stewart Gardner Museum.

A charming, ambitious, headstrong woman with money, who

adored art of all kinds and enjoyed conversation, flirtation, dancing, and singing—all of which she did beautifully—Gardner created a climate for art that did not exist before she arrived on the arm of her proper Bostonian husband, John Gardner. She took the city (and a lot of now priceless European art) by storm, a storm of charm, scandal, ambition, and will to build a distinctive collection. Yet it was her human touch, her genius and generosity as patron, mentor, and friend that constitute her larger legacy. By visiting artists in their studios, inviting them to her home, encouraging them, and introducing them to each other and to her society friends, she made "cold roast Boston" hospitable to art. She was the city's cultural ambassador, in the vanguard of those who would turn Puritan-Victorian-provincial Boston into what civic leaders today call a world-class city.

Isabella Stewart Gardner—called "Belle" in her New York incarnation, "Mrs. Jack" once married to Jack Gardner and in Boston—always loved beauty. In his compelling biography *The Art of Scandal*—at once deliciously dishy and erudite—historian Douglass Shand-Tucci writes that during her girlhood, after seeing the Poldi Pezzoli Museum in Milan, Belle confided in a friend that if she ever had money of her own, she would like to have a house like the one in Milan, "filled with beautiful pictures and objects of art, for people to come and enjoy."

Shand-Tucci follows Gardner's evolution from a privileged, high-spirited New York girl (albeit the granddaughter of a tavern owner) to a society woman who assumed and shirked roles as it suited her, educated herself in art and architectural history, in philosophies of aesthetics and religion, and with growing confidence and the counsel of intimate advisors, followed her own tastes. During late middle age, after her husband's death, Gardner assembled a highly original, world-class collection—bringing Italian art and architecture to America *by designing a particular building for particular art*—during a time in nineteenth-century Boston that prescribed for widows a retreat into needlecraft. Gardner's only interest in needlecraft was in selecting opulent textiles that she could commission into figure-flattering clothing, artfully arrange on chairs, or turn into draperies and spreads, including those seductively lapped under trays on a bed.

Events in Gardner's early married life, including the death of her only child, Jackie, at age two, and the suicide of a beloved nephew, who was homosexual, resulted in a paralyzing depression, which eventually—with hard work, that of self-transformation—led to a created, creative life. Her husband, Jack Gardner, who may not have been on her wave length, but who loved and treasured her, brought Isabella to Europe to try to alleviate her depression, to bring her back to life. Italy, in particular, spoke to her, with its beauty, sensuality, spirituality, and the Italian way of using emotion and feeling to "go" somewhere—to reach out, create, and express, rather than to repress, regret, and suffer. Shand-Tucci suggests that Gardner's profound depression may have resulted not only from the loss of her child, but from her repressed sense of unsuitability for the limitations of motherhood. What she saw and experienced in Italy may have planted a seed for a different way of life, a way of being a woman alone at the turn of the century.

She started to buy what caught her eye. (She was not wholly reliant on her husband's money, as she had inherited some of her own.) She began to study art through reading and touring, sketching and keeping a journal, studying with Charles Eliot Norton, professor of art history at Harvard, and by cultivating relationships with artists. During the second half of the nineteenth century, she developed a salon in her Beacon Street home and later in Brookline.

When her husband died at age sixty-one, Gardner channeled her sadness and his money into building Fenway Court, a showplace of art, architecture, and antiques that became a gathering place for artists, writers, and composers from all over the world. It was a combination Museum of Fine Arts, Boston Symphony, Athenaeum reading room, Italian loggia, Victorian conservatory,

Italian courtyard, medieval cloister, Anglican chapel, Buddhist temple, Ritz dining room, proper tea room, improper wine bar, art center, and what we would today call a niche hotel. Over decades she developed a new society in Boston that crossed Brahmin and Bohemian lines, and created a collection that encompassed art of all periods and types.

On the Brahmin end of the scale, she was of high social standing, active in the Anglican Church, well-off, and known for her great parties. She was sufficiently scandalous to attract useful attention. On the Bohemian end of the scale, she had a passionate love of beauty, art, and artists; independence of spirit and mind; and boundless enthusiasms for ideas and their synthesis. Gardner bridged these worlds, guiding travelers back and forth.

She was the first female American collector who designed, built, and curated her own museum. And she wasn't even a royal! She was the first American impresaria of the arts, as only someone with her combination of worldly and innate qualities could have been, and among the first to practice in the new fields of landscape architecture and interior design.

She lived between two centuries, spanning the Victorian to the Modern age. She began as a married woman in Boston society and became a relatively young widow, and also a kind of artist herself, designing a site-specific collection in a Venetian palace. Shand-Tucci says Gardner even originated the contemporary concept of the museum as cultural center, a setting for art, performance, lectures, dining, education, and social and intellectual exchange.

Isabella Stewart Gardner lived as a woman who could charm and beguile, command and demand, assign and consign—because of her money, knowledge, chutzpah, and vast array of friends. Artists, composers, writers, poets, actors, singers, dancers. Bankers, investors, philanthropists. Politicians, prelates, royals. Painter John Singer Sargent was a buddy, as were composer Charles Loeffler, novelist Henry James, and Museum of Fine Arts curator (and also artist and poet) Okakura Kakuzo. Julia Ward Howe, the great feminist and reformer, was a mentor and cherished friend, as was her daughter, the writer Maude Howe Elliott.

The Isabella Stewart Gardner Museum at Fenway Court, which officially opened in 1903—in a gala New Year's Night bash—is today a one-of-a-kind pleasure palace of painting, sculpture, and decorative arts, and also a visual, sensual ode to the joys of created atmosphere. The palace (sedate in exterior, lush within) contains a formal courtyard garden (with seasonal plantings in ceramic urns and pots) and a small, perfect universe of light-diffused rooms decorated with and exhibiting furniture and art. Today, as in Gardner's lifetime, concerts and theater pieces are staged in a salon.

As important as the museum, Gardner herself, more than any individual in the nineteenth or twentieth century, made Boston safe for art, art making, and artists. Muse, mentor, patron, collector, and curator, she encouraged those in her circle, bringing them together for hundreds of soirees, shows, and gatherings over the decades of her life—and after! She was a social activist in a distinctly modern, American sense. Her closest associates, colleagues, and friends were homosexuals, and her artistic advisor, Bernard Berenson, was a Lithuanian Jew who grew up in Boston's West End (though granted, a Lithuanian Jew who went to Harvard). Deeply religious, albeit self-styled, Gardner bought the land in Cambridge for the Cowley Fathers to build their monastery and funded the completion of St. Augustine's Church on Beacon Hill, where in 1894 the first African American priest was ordained in the Episcopal Church in Boston.

She became a modern woman, a person who uses her life. Her interest in art caused her to seek the company and knowledge of creative people in many fields. These colleagues became lifelong friends, loving and loyal till the end, setting up musicales in her sickroom, sending over drawings and poetry, and even warm dinners from Boston hotels (risotto was a favorite). Over many years, they had all come to Fenway Court, not merely for luncheon or tea, but for intellectual exchange and entertainment, and to live in the Bostonian-Venetian apartments at the palace, sometimes for months at a time.

These associations are part of Gardner's legacy—weavings of people, cultures, and ideas as precious as any antique paisley draped on a four-hundred-year-old Baroque chair, or a medieval tapestry on an artfully colored wall. Gardner's rare combination—patron, mentor, and devoted friend, as well as impresaria, diplomat, and cultural ambassador—was a giant step in puritanical Boston. *C'est mon plaisir*, it says over the doorway of Fenway Court. Her credo, "It is my pleasure," powerful during her life and now, is an antidote to the parsimonious, the spartan, and the narrow-minded, to the boring and pointlessly chaste. It was and is an explanation, a theme, a coda—above all, an invitation.

RECREATION
& CELEBRATION

MADAM ALICE

❦ 1 6 7 2 ❦

Even the Puritans had sex, and for pleasure, not just procreation. Rude lodgings, lumpy beds, and the threat of public lashing didn't stop it. Desire burned well before the advent of Victoria's Secret, the Miracle Bra, and the Personals. Boston's seventeenth-century Puritans did everything that we do, but less (they had more work and fewer opportunities), and with more guilt. They had sex indoors, outdoors, and in between—on dark porches, on bundles of straw in barns, and in carts pulled into the woods.

Their activities (discernible via punishments meted out) are memorialized in documents such as the Records of the Suffolk County Court, where in precise and learned writing are chronicled an array of lively goings-on. Mention is made of a Mrs. John Gill "in company of severall men at the house of Arthur Keyne, drinking and dancing." A man is found guilty of licentious behavior toward his maid, and many of "making Sute to some maids or women in order to marriage," despite having wives in London. Married women, no longer young, are observed "sitting in other mens Laps with theire Armes about theire Necks."

From the start, there were prostitutes in the colonies working on their own. But Alice Thomas was the first to organize a house of prostitution—the first managerial woman in the sex trade. Once a proper Puritan wife, she was widowed, ran a shop, and had to enter the ancient profession. She was a tough customer, the first American madam on record, and hard not to admire. Her clientele and experience are a Rorschach of life and mores in seventeenth-century Boston.

Information about Madam Alice is scarce, but exhaustively gathered and well described in Peter F. Stevens's compilation, *Notorious and Notable New Englanders*. The author draws us into the community Thomas occupied and, even without having much first-hand testimony of Thomas's personality and character, it is no stretch of the reader's imagination to regard her as capable, strong, and self-reliant, not to mention enterprising and adaptable. From Stevens's accounting and the ancient court records, Thomas emerges as "a common baud," but also one sharp cookie, though too self-styled and irreverent to escape punishment.

While her business was against the law, and the activities fostered in her establishment defied Puritan religious practice, it was her violation of social norms—her big-time breach of etiquette in arranging sex for a married merchant-prince—that inflamed the fires of vengeance. She conducted her business out of her home, administrating the favors-for-hire of young women, many of them servants who hoped to earn money to buy their way out of servitude. Her specialty was not unknown, nor was her address, nor the well-trod path to her home, nor the variety of available pleasures, and their cost. (Even the frugal Puritans would shell out for a good time, a thought not altogether discouraging.)

What we would today call a nasty divorce case proved to be her undoing: the meltdown of the marriage between merchant Edward Naylor and his wife, Nanny. Nanny, née Wheelwright, belonged to a powerful, wealthy family, determined to drag the once estimable Edward through the dirt. He was observed in the act with a servant woman, Mary Moore, by Puritan John Anibal, who had apparently found it pure to stand outside Madam Alice's house and snoop through the window. "I have often seen Mary Moore and

Mr. Naylor at the Widow Thomas' house together," Anibal reported to the General Court in 1671.

That Madam Alice's establishment had given comfort to a citizen such as Naylor, tempting him, besmirching him, dragging him to moral ruin—which, in the Puritan sense, stained the whole community—was what enraged the would-be peeping populace. Alice was hauled before the court and charged as a "common Baud," arraigned and placed in Boston Prison. This horrid enclosure, the first prison in New England, had been built in 1637 on the orders of John Winthrop.

To understand the role, censure, and punishment of Alice Thomas, it helps to understand a bit about the community she served. Seventeenth-century Puritan Boston was remarkable for the amount of involvement of the government, which is to say the clergy, in the private lives of citizens. In the other British colonies, common-law marriage was not unusual, nor was it frowned upon. But in Puritan Boston, where marriage had to be sanctioned by a civil or religious ceremony, both adultery and bastardy were punishable. Threats to the contract of marriage were taken seriously.

Browsing the court records (which can be done at the Boston Public Library), it is tempting to imagine this early Boston as a Puritan Peyton Place, but this was not the case, according to Zechariah Chafee Jr., who wrote a fine introduction to the collected *Records of the Suffolk County Court.* The crimes on the books are probably a high percentage of all the crimes committed, suggests Chafee. There was no police force; neighbors knew one another, and so ordinary citizens "patrolled" their communities. Madam Alice was turned in, and her john—actually the john of working girl Mary Moore—was revealed because a townsperson spied, observed, and told.

In January 1672, Alice Thomas and her brothel were busted. She was brought before the magistrates and charged with an accumulated host of crimes, enumerated by the court records: "aiding and abetting theft by buying and concealing stolen goods; frequent secret and unseasonable entertainment in her house to lewd, lascivious and notorious persons of both sexes, giving them opportunity to commit carnal wickedness; selling drink on the Sabbath."

The punishments she was subjected to—mental and physical abuse, really—are painful to read and fearful to contemplate. First she was dragged to the gallows with a noose about her neck, townspeople gawking, as magistrates debated whether to hang her for thievery.

She was then stripped above the waist—this in a Puritanical society, and in the terrible winter of 1672—roughly tied to an oxcart, where thirty-nine "stripes" (lashes), probably with a switch, were cut into her back. Freezing, bleeding, half-nude, she was dragged by the cart and paraded through town. Curses, jeers, and rotted, stinking vegetables were hurled upon her. She was then returned to jail, ordered to serve a sentence until October, and banished from Boston forever.

Somehow—and here the remaining sketches of Alice Thomas's story grow faint, as though bleached by time—she managed to not only reemerge, but to reestablish herself. Did she achieve rehabilitation by good works, or good behavior in prison, or might she even have impressed the magistrates by her self-reliance and pluck? Thomas never revealed her clients, which might imply a professional integrity, or a triumphant deviousness, or keeping to her own code of conduct. What is known is that she made good money in her trade, kept it, healed sufficiently from her lacerations to function, and made monetary contributions to the community. From 1673 to 1676, according to author Stevens, Thomas made donations to Boston's construction of a much-needed harbor seawall and common buildings. In July 1676, the decree to banish her forever from Boston was rescinded. To the chagrin of the Wheelwrights and others who wished her ill, Alice Thomas returned to Boston and lived quietly and independently on the money she had earned.

She had ventured, prospered, risked, and lost almost everything. A careful businesswoman, she had kept her own counsel and
never invested more than she could afford to lose. Alice Thomas
regained her dignity, made a place for herself, outlived many of her
enemies, and, for the rest of her life, paid her own way.

HOTEL

❦ 1 8 2 9 ❧

To read of Tremont House is to feel a twinge of longing, though the gracious hostelry was built almost two centuries ago. How fine it would be to arrive in a carriage, passing Boston Common, to then take in the sight of this softly gleaming white hall, almost like an athenaeum—with portico and colonnade—so in keeping with the dignity, scale, and verve of the city.

To see a photograph of the now vanquished palace of urbanity, sophistication, cleanliness, and romance—the first modern hotel, built 1828–1829 on Tremont Street—is to feel desire to luxuriate in its well-appointed rooms, to dress carefully for dinner, to observe the glitterati in the lobby. (But the lobby was not yet called a lobby. Hotels of the day had "offices." They did not have bellhops, either, but "rotunda men," robust young men, in from the provinces.)

To learn, even in our era of consumer satiation, of the services and offerings of Tremont House brings on appetite and thirst, a wish to order a sumptuous meal that magically appears. The Tremont was the first American hotel to offer à la carte dining, whereby one could order from a menu, as opposed to supping on whatever the innkeeper presented. The main dining room, seventy feet long and thirty-one feet wide, with gleaming marble floors, could serve two hundred swells at a sitting. It was the first hotel to rent individual, private rooms; one no longer need double up with a stranger, as though traveling on the rough frontier. It was the first hotel to

❦ 137 ❧

provide a washbowl, a water pitcher, free soap, and a high level of personal service. Never before had there been different locks on the doors of each room and tagged room keys. The identifying tag now affixed to every hotel and motel key in America—from the Ritz to the Ramada—originated with Tremont House, as did the custom of leaving one's room key at the desk.

Above all, to fully enter a fantasy reprise of Tremont House is to anticipate and crave a bath, to remove oneself from society—from the pressures of conversation, evaluation, and stiff, unyielding stays sticking into one's rib cage—and to recline, relax, and soak. To submerge belly and pelvis, shoulders and neck, in deliciously hot water that stays hot! To be surrounded by shiny copper, to view thick Turkish towels waiting, to play with a smooth, oval cake of finely milled soap. No charge for the soap, precious soap, included in the cost of two dollars per night.

Boston's first luxury hotel had the first bathtubs with hot running water—no small thing if, more generally, bathing forced you to scrunch into a basin with water that had been heated on the stove, carried in pitchers, and poured into the tub, which then had to be emptied—unless its contents were used to bathe the next, less fortunate, family member in line. Even at Tremont House, some plumbing problems remained unsolved. The bathtubs were in the basement, as it was still not known how to move water up into rooms. Each of the commodious bathtubs, some copper, some tin, was fitted with cold running water and a small gas furnace at one end. The water flowed into the tub, then circulated until heated to the bather's preference. As there was no municipal water supply, the Tremont, like other buildings with running water, had its own. Clean water for bathing—and also for the kitchen and laundry—started its journey to hotel guests in a metal storage tank on the roof of the hotel (the recently invented steam pump got it up there), and a simple system carried the used water into the sewer.

The creator of this pleasure palace was twenty-eight-year-old Isaiah Rogers, born in Marshfield, Massachusetts. He descended from John Rogers, who had arrived in 1647. Tremont Place launched young Rogers's career and promoted the glories of Quincy granite: its sheen when polished, its grandeur when rugged, its unsurpassed strength. Rogers (1800–1869) became a distinguished, much sought-after architect, creating monumental granite buildings, including banks, hotels, churches, mansions, and theaters. Early in his career, he worked mainly in the Greek Revival style, and later in the less monumental, more Italianate form. Immediately after Tremont House, Rogers designed other modern hotels, including New York's Hotel Astor. Of course, New York had to outdo Boston, and the Astor is fancier than the Tremont—two stories higher (the Tremont rose three and one-half stories at the front, with four-story wings at each end), with twice as many rooms, and costing $400,000 to construct—$100,00 more than Tremont House.

Several of Isaiah Rogers's grand buildings in Boston survive, though most have been altered. Commercial Wharf at 84 Atlantic Avenue (1832–1834) uses Quincy granite beautifully. The building's lower walls are rustic, rough-hewn, while its lintels and string course are smooth-cut, in contrast. In the same harbor neighborhood, Rogers's monumental Custom House Block (1845–1847) also remains—on Long Wharf, off Atlantic Avenue—a massive, long building with a graceful center arch that again showcases Quincy granite. (Nathaniel Hawthorne's "day job" was in this Block; he worked as a customs inspector.)

The amenities of Tremont House that stimulate our desires even now are what amazed hotel guests of the early nineteenth century, and made Bostonians proud. The creation of such a hotel was a great civic undertaking and event; that Boston would warrant such a fine hotel had been the talk of the town for several years. The cornerstone was laid in a gala ceremony on the Fourth of July, 1828, the fifty-second anniversary of Independence. Levi Lincoln was governor, and Josiah Quincy, that great promoter and promulgator of a world-class city, was mayor of Boston.

Then again, the hotel's luxury—and staff trained to fuss over guests—was not for every taste. It horrified some diehards. A lodging that encouraged choices in dining could only repulse those raised to eat what was put before them, preferably simple fare, and the idea of hot running water in a large bathtub—the confusion of pleasure and luxury with cleanliness!—distressed those accustomed to bathing sparingly, part by part. (During the nineteenth century, all sorts of bathing appliances, such as anatomically shaped water holders—bidets for different parts, really—were available, each designed to cleanse a different area. No need for total immersion!) In *Lost Boston*, architecture critic Jane Holtz Kay quotes Julia Ward Howe's daughter on her grandfather's first and only visit to the Tremont: "Grandfather—a Puritan of the Puritans—fled from it in terror."

Hotels are small cities. Everything that can happen to men and women happens, room by room. All needs are experienced and consummated. In the public rooms, guests observe each other, meet, and form associations. The atmosphere and attitude of the hotel—its architecture and décor, policies and practices—form a small society. The hotel also mirrors its larger setting, and can exemplify its best qualities. This was particularly true of the Tremont in an era when hotels were rare, and a grand hotel of its nature singular.

During much of Boston's nineteenth century—an era of formation and consolidation, invention and enterprise—Tremont House was an architectural presence, physical and symbolic, extending Boston into the national and international realm. Famous guests ranged from presidents Martin Van Buren and John Tyler to writers Charles Dickens and William Makepeace Thackeray. In *Lost Boston*, Jane Holtz Kay quotes an amazed Dickens, observing of the Tremont, "It has more galleries, colonnades and piazzas and passages than I can remember or a reader would believe."

Sadly, almost inconsolably for those of use who long for its galleries and colonnades, piazzas and passages—and the salubrious, steamy basement baths—the great hotel was demolished in 1895.

The site of Tremont House—across from the Granary Burying Ground and Athenaeum, and steps from old Boston City Hall—is now occupied by the Parker House. This is also a hotel worthy of affection, if not veneration, and is the birthplace of Boston cream pie (see page 173). The dessert, which can assuage all longing, is still prepared in the present hotel. Like the ghost of Tremont House, the pie-that-is-cake is elegant but friendly, luxurious yet comfy, and it carries a deliciously conflicting chaste/come-hither quality. Like the service at Tremont House, the pie is American and French, and can be ordered à la carte.

One could arrange for oneself a brief, eccentric tour of Boston, an homage of sorts: Stay the night at the Parker House, take a long, delicious bath (free soap), and order Boston cream pie à la carte. As you leave the hotel the next morning, look about. You will see many vestiges of Boston that travelers gazed upon and admired during the long life of America's first modern hotel, a great hotel, Tremont House, 1829–1895.

YMCA

Imagine what life was like for young men—boys, really—alone and adrift in Boston during the 1850s, away from home for the first time. Today, even if we profess to "never having been anywhere," we cannot escape knowledge of other places, other kinds of people, other ways of living. (And who among us has grown up from childhood to late teens in an unchanging atmosphere?) But during the mid-nineteenth century, the boys arriving in Boston had generally spent their lives in one place, often a rural place, seeing the same sights and the same group of people—perhaps under a score of relatives and near neighbors—for all of their lives. If they had worked, it was most likely at all phases of an operation— farming, or building, or helping with a family enterprise—from start to finish to make something.

Suddenly, they were on their own in a boarding house, or sleeping over a shop, or even sharing a bed with a stranger, making their way through cramped and clamorous streets, working at repetitive jobs, the same task or maneuver over and over for most of the day (or night), with no connection with what they produced. No one guided, informed, or encouraged them. In some cases, there was no conversation at all.

Many young men got into trouble, wasting their money on coarse pleasures and frittering away leisure time. They got drunk, sick, and lost in the streets. They were beaten and robbed. Those who could not adjust were, above all, unhappy. To put it in modern terms, many of these boys—sailors, laborers, factory workers —were clueless about their environment, frightened, and estranged. They felt themselves unmoored, their surroundings jarring and surreal. They were painfully lonely.

It was the same in industrial cities all over Europe. Commerce was booming, and railroads delivered workers from the provinces, cars of young men like grain, waiting to be processed. A sane, humane solution arose in England in 1844. A fellow named George Williams (who would one day become Sir George Williams), age twenty-two, a farm boy turned draper in a London shop, and a group of eleven friends—drapers all—started a club where they prayed together, studied the Bible, and reached out to offer their fellowship to other young men, particularly newcomers to London. They wanted their club, the Young Men's Christian Association (YMCA, or Y for short), to serve the social, religious, and educational needs of young men.

Then, because of a fortuitous turn of events, two men were major players in getting a Y to happen in Boston. One merely wrote a letter; the other—with the perfect middle name, Valentine—was the galvanizing force. Boston's was the first YMCA in the United States, and led to the movement all over the country. The constitution for the Boston YMCA, which emphasized government by a community, became the model for all the American YMCAs, just as the Massachusetts Constitution (see page 201) became the model for the U.S. Constitution.

Captain Thomas Valentine Sullivan (1800–1859), a kindly Boston-born seaman turned missionary, founded the Boston Y. But he couldn't have done it without George Van Derlip, a young New Yorker studying in Edinburgh, who chanced to visit the YMCA in London—*the* YMCA, founded by George Williams—and wrote about it during the summer of 1850. What Van Derlip found— a clean, welcoming, wholesome place, an environment that was friendly but not fancy, Christian but not narrowly sectarian, and that provided books, refreshment, and fellowship—became the model for the Boston Y.

Van Derlip submitted an essay, in the form of a letter, about his discovery to *Christian Watchman and Reflector*, a Baptist weekly in Boston. The editor put the well-traveled piece in a pile and, for reasons unknown—though possibly because of the mounting problem of rowdy young men in the city—published it a year later, in October 1851. Captain Thomas Valentine Sullivan, by then in Boston, trawling the misery of its wharves, tenements, and streets, read the front-page essay and determined to found a YMCA in Boston. Years later, Van Derlip would modestly refer to himself as the "innocent cause" of the founding of the American YMCA.

During the early years of the YMCA's existence in London, Captain Sullivan had been living the only life he had ever known, that of a seafarer. He had shipped out as a boy of nineteen, joining a sealing expedition to the Antarctic. In his midthirties, mentally and physically exhausted by shipwrecks, piracy, and illness, the battered sailor experienced a religious conversion and helped to found the American Bethel Society in Buffalo, New York, where he provided comfort for sailors in the Great Lakes area. (Though of Irish descent, Sullivan was raised a Baptist.) He was a warm and understanding man, a good listener with an open heart. Sage, encouraging, inspirational, he went from ship to ship in the inland waters—a kind of spiritual medic—and became a Marine Missionary, serving Christians on rivers, canals, and lakes, and converting many a man to Christianity. He described his work as "social religion"—locating the men who did not attend services, encouraging them to learn more skills, and even establishing a marine lending library.

Sullivan continued to love his birthplace, and returned to Boston in 1847. Trained by his experience ministering to sailors literally and figuratively tossed about, he noticed the misery of the young men, mainly newcomers, on Boston's streets. The city had greatly changed since his boyhood. Pickpockets and beggars were everywhere, as were filth, stench, and crowded tenements. Pleasures of the flesh were openly peddled. Sullivan felt a kinship with the young men he saw wandering about, cowed by the city, as well as a deep compassion and concern. Many of the boys were sailors, as he had been. Many were immigrants, like his own Irish grandfather.

Sullivan, both a spiritual and a worldly man, understood the power of Boston's many churches, their alliance with prominent businessmen, and the way community and organization could move decisions forward quickly.

The city's Protestant evangelical churches—Methodist, Baptist, Episcopal, Congregational—expressed immediate interest in Sullivan's proposal for a YMCA. Meetings began in November 1851; deliberations focused on the social concerns of the day—the influx of foreigners, the rivalry among religions, and the move from narrow sectarianism toward openness. The first YMCA in America was founded a few days after Christmas: December 29, 1851, in the Spring Lane Chapel of Old South Church. (Old South had remained staunchly Congregational in a tide of Congregational churches turned Unitarian. Boston's urban Protestants were mainly Unitarian, while the young men arriving from the provinces were generally evangelical Protestants.)

By January the enthusiastic group of clergymen, Christian businessmen, and community leaders had rented space in a handsome new granite building on Washington Street at the corner of Summer, the site of today's Macy's—a suite of attractive rooms for lectures, socializing, and reading. When the YMCA facility opened in March 1852, the *Boston Daily Journal* ran a long, laudatory front-page story, reporting "a large attendance" at the opening bash and "good taste" in décor: "Both these rooms are brilliantly lighted with gas, and present when lighted up an air of comfort and neatness which cannot fail to make them a pleasant place of resort for the members." Over six hundred men attended the gala. Addressing them, the Reverend Dr. Lyman Beecher declared, "I always felt sure the millennium would come, but never so sure of it as now."

This Young Men's Christian Association in the heart of Boston— flooded by light during the day (commanding the top floor of a corner building) and cozy at night with state-of-the-art gas lamps—was not *like* a home, it *was* a home. The YMCA provided friendship and support, gracious décor and refreshments ("sold at cost"), prayer meetings, Bible study, lectures, social events, and a peaceful, quiet reading room, stocked with newspapers from all over New England, so that a young man might find something familiar, a touch of home.

Though other YMCAs soon opened across the U.S.—New York, Washington, D.C., Buffalo, Detroit, Springfield, Massachusetts, all in rapid succession—it was not accidental that the first association opened in Boston. Nor was it only a matter of the instrumental Captain Sullivan's return to his birthplace. Boston, a port city, railroad hub, and commercial and manufacturing center, was a magnet for young men searching for economic opportunity, including tens of thousands of immigrants. According to a YMCA history from the turn of the last century, the rural population of New England increased by just 12,000 during the years 1850 to 1890, while the urban population grew from 136,831 to 1,200,000 —one hundred times the rural population increase. In 1850, almost half of Boston's population of 138,788 was foreign-born or of foreign parentage, according to the 1850 state census.

Boston had a strong Christian underpinning and what might be called a useful tension among its several branches of Christianity. During the early nineteenth century, relations were strained not only between the Catholic and Protestant churches—the tide of Catholics rising dramatically with the influx of Irish immigrants— but among different Protestant denominations. The organization of a Christian club for young men enabled evangelical Christians (Congregational, Baptist, Methodist, Episcopal) to unite. Unitarians, considered too liberal, were not welcomed by the Y's founders, though eventually everyone was welcome at the YMCA, one of its hallmarks in American culture.

In addition, Boston had a strong tradition of joining, of participation in organizations. Despite the image of the reclusive, laconic, self-sufficient Yankee, this was never a city of lone rangers, but a society of associations—clubs, fraternal orders, community groups, town meetings, trade associations, church fellowships, charitable organizations, and literary societies. No less influential a figure than seventeenth-century clergyman Cotton Mather had written in a letter to his son, Samuel:

> I will further inform you, my son, that a singular advantage to me while I was thus a lad, was my acquaintance with, and relation to, a society of young men in our town, who met every evening after ye Lord's day, for ye services of religion.... As ye Lord made poor me to be a little useful unto these, and other meetings of young people, in my youth, so he made these meetings very useful unto me.

In the years to come—a century and a half—the Y became enmeshed in the lives of hundreds of thousands of Bostonians, and millions of children and adults across the U.S. The association led to the founding of a university. By the 1880s, hundreds of students were enrolling in evening classes at the Boston YMCA, studying vocational subjects ranging from bookkeeping to electricity, diverse foreign languages, and liberal arts such as singing, elocution, and grammar. In 1896 these courses were bundled into the Evening Institute for Young Men, and Frank Palmer Speare was hired as director. By the turn of the century, these classes had become Northeastern University.

Several sports are closely associated with the American YMCA. Basketball was invented by James Naismith, a physical education teacher at Springfield College in Springfield, Massachusetts—at the time, the International YMCA training school—in 1891. Volleyball was invented by instructor William Morgan at the Holyoke, Massachusetts, YMCA in 1895.

Today, most YMCAs in the U.S. have become state-of-the-art fitness facilities, the better to compete with commercial gyms. But in the old industrial cities such as Boston, some of the brick-and-stone, karma-infused, early twentieth-century "Y buildings" remain. Boston's Central Branch on Huntington Avenue, constructed in 1912, the grandfather of all Massachusetts Ys, has been modernized, and is today an impressive, well-equipped facility. Members range from students to neighborhood people to players in the Boston Symphony Orchestra, exercising to keep their playing arms flexible and strong. But for the visitor alert to clues of karma, the pool still contains some original tile (almost one hundred years old!), and the original gym is a shrine to push-ups past. The Central Branch is a few short blocks from Northeastern University, a descendent of Boston's 150-year-old YMCA.

Even today, should a young man come from afar, he would find a welcome at the Boston Y. In addition to the glass-and-steel fitness facility and the whir and the clank of its equipment, a traveler would find safe lodging, nourishing food, and someone to talk with who will listen carefully. Kind guidance is the main legacy of Captain Thomas Valentine Sullivan, the Boston seaman and lay missionary who became known as Reverend Sullivan.

WORLD SERIES

❦ 1 9 0 3 ❧

Boston, *Autumn, 1903, exterior Roxbury bar.*
American flags snapping in the breeze, bunting draped above windows, leaves swirling in the air. (The bar is decorated for Boston's triumph in the first World Series.) *Interior:* red, white, and blue streamers decorate the high ceilings and long bar, which the barkeep, a small man with a waxed handlebar moustache, makes a great show of wiping down—flourishing and twirling his towel. People are wall-to-wall, holding up pints of beer and shots of rye, singing "Tessie," a popular song of the day.

Dissolve to: Boston, Winter, 1900, exterior Roxbury bar. Snow swirling. *Interior:* paneled wooden bar, light fixtures made from baseball bats; working men, sleeves rolled—laughing, talking, drinking, arriving after factory shifts. The sun goes down as the guys order whiskey and beer, and the occasional meat and potatoes-platter, which the regulars watch being served. "Wife throw you out, Joe? How come she waited so long?" Wisecracking politicians, reporters, cops pepper the scene.

The talk at the brick corner hangout, number 940 Columbus Avenue, is all-baseball-all-the-time, including all winter long. Though officially M. T. McGreevey & Co., the saloon is called Third Base, "the last stop before home." The regulars descend in droves after baseball games at the nearby South End Grounds, and during the rest of the year—to reminisce, banter, and anticipate next season.

It's fun to picture the legions of guys jawboning and pulling down shots at Nuf Ced's Third Base on Columbus Avenue, circa 1900, dreaming of glories for the home team. If only wishing made it so. Then again, a form of wishing—call it "mobilized wishing" —eventually did make it so. Within a year, the fans' loyalty— fierce, heartfelt, ancestral—would enable them to win over the owner of a new baseball league, convincing him to build a ballfield in Boston.

Michael McGreevey, the larger-than-life saloon owner—Godfather of the World Series from the fans' perspective—was known as Nuf Ced for his decisive method of ending an argument. The guy—small, but pumped—would slam his fist on the wooden bar and bellow, "'Nuf said!" He was five feet tall, sported an outrageous handlebar moustache, and appeared in a natty dark suit, high white collar, and black bowler hat on public occasions. A fun-loving Irish barkeep turned brilliant, avocational publicist, Nuf Ced led the Royal Rooters, the antic yet foxy Boston brigade who campaigned mightily for the home team.

McGreevey's Third Base saloon was Faneuil Hall with whisky. At the meeting house for the baseball faithful, the great sports debates of the day took place, as well as those political. These were guys with the gift of the gab—savvy workers, politicians, newspaper reporters (who gave as much ink to the Rooters as the players), and also the athletes themselves. If you cared about baseball, you were at Third Base. Estimates have it that 15 percent of the guys in the stands were Nuf Ced boyos.

Baseball's appeal in Boston was not new. Many sports scholars opine that it began here. In precolonial New England, baseball was evolving from different sorts of English ballgames, according to Glenn Stout and Richard A. Johnson, authors of *Red Sox Century: The Definitive History of Baseball's Most Storied Franchise.* During the 1600s, it was devilishly popular in Plymouth colony, where an alarmed Governor Bradford banned it. Ballplayers were turning into idlers, worried the governor, their livestock wandering off to

frolic (a double sin, both man and beast having fun), fruit wither-
ing on the vine, shocks of corn mottling.

Despite the governor's decree, the game survived for two hundred years into the mid-nineteenth century. By the late nineteenth
century, Boston was Baseball Central. When the National League
organized in 1876, the Boston team joined. Writers Glenn Stout
and Richard Johnson say the team was "a terror," playing like
men possessed, winning a dozen pennants over the next quarter-
century. They also won the fans' devotion, especially the guys who
gathered at Third Base. Boston fans accustomed themselves to
winning, which is why doldrums at century's end made the Royal
Rooters queasy.

By 1900, there was major grousing. Boston had not only lost
the pennant, but teams were chaffing under the system of syndi-
cate baseball. Gambling was rampant, a well-established part of the
game, but there were limits (imposed by the fans' sense of fair play,
if not by the legality of the practice). It disgusted fans to discover
that owners bet against their own players, and bought shares in
other syndicates—gestures of contempt for the athletes and a dis-
honoring of the game. Baseball was a working-class sport. Fans'
sympathies were with the players, who were often Irish lads, just
as most of the early fans were.

Team members frequented McGreevey's and bellyached about
the bosses. The players were treated badly by the National League
—as "cash cows" to use modern parlance. They particularly re-
sented the reserve clause in their universal contract, and the salary
cap of $2,400 per year. The working stiffs at McGreevey's took this
on as their grievance, too. They were mainly union men—factory
workers and men in the trades, campaigning for better conditions.
They viewed the League as just another greedy boss.

Meanwhile, far from Roxbury, a tall, vigorous Ohio native, who
had played baseball as a kid, was looking to start something new—
bigger, better baseball.

Byron Bancroft "Ban" Johnson was a smart guy who couldn't

stay behind a desk. He spent time in law school, then went to work for the *Cincinnati Commercial Gazette* as a sportswriter, and then became a sports editor. Before long, Ban Johnson knew everybody connected with baseball in the Midwest. He left journalism and became president of a small Midwestern group, the Western League, a mere fly in the eye of the National League, which included the Boston team. Late nineteenth-century baseball was rowdy—fighting, cursing, and carousing in the stands, and unsportsmanlike behavior in the field. Johnson envisioned a classier game, where ladies might attend; he cleaned it up, outlawed roughhousing, and treated his players decently. His league became competitive in cities such as Milwaukee, Detroit, and Indianapolis, which the National League had ignored.

In 1899, the increasingly confident Johnson changed his group's name from the Western to the American League—dropping the regional association—and began to invade the turf of the National League, moving into Baltimore and Washington. He tried to organize syndicated baseball into a two-league organization. The National League rebuffed him—told him "he could go to hell," according to *Red Sox Century*—and Johnson decided to go his own way. He looked toward Boston, a baseball town with a strong fan base, a losing team, and growing discontent. The Nationals were entrenched in Boston, but it was worth a shot.

Ban Johnson sent Connie Mack, a player turned manager in the American League, to test the waters in Boston and investigate a site for a ballpark. Cool move. Though Mack lived in Philadelphia, he'd grown up in East Brookfield, Massachusetts, and was popular in Boston, remembered both as a player and a native son. Nuf Ced and the guys were ready; they hadn't dubbed themselves Royal for nothing. The Rooters gave Connie Mack a splashy welcome. By the second day, Mack, assisted by the Rooters, had found a place for a ballpark.

The Huntington Avenue Grounds were pretty close to perfect. They were near the New Haven Railroad yard—not useful for housing, and therefore affordable—in the middle of a working-class neighborhood of Irish immigrants. The land was near the South End Grounds, too, the park of the National League, a nice, neat slap in the face. It was also close to Third Base, consolidating operations.

Poof!—or so it seemed—a ballpark went up on Huntington Avenue. "How Can You Get Home Without Reaching 3rd Base? Nuf Ced," read a huge ad on the wall of the spanking-new wooden ballpark, described by Roger I. Abrams in *The First World Series and Baseball Fanatics of 1903*. The Royal Rooters, many of them shift workers, dropped by during its construction during all times of the day and night.

Reserve clauses be damned—Johnson, who had tried to enlist the support of the National League and been dissed, went after the best on the National League's Boston team. None of the players had any great loyalty to their arrogant bosses, but they had signed contracts. Johnson zeroed in on third baseman Jimmy Collins, a star whose move would attract others, offering Collins four thousand dollars. In February 1901, Collins jumped teams. (*Red Sox Century* quotes him: "I like to play baseball, but this is a business with me....I look out for James J. Collins.") Other Boston players followed, and stars from afar, including legendary pitcher Cy Young, playing for St. Louis. Young's fee: $3,500.

(Today, Northeastern University is on the site of the Huntington Avenue playing field. A life-sized bronze statue of Cy Young —the agile pitcher winding up—is in the shrubbery just outside the university president's office, near the old pitcher's mound.)

American League baseball took off big-time in Boston. Great games were being played, the fans were happy, and Nuf Ced's joint was jumpin'. The Bostons got stronger and stronger. The 1903 season yielded two talented, spirited, victorious teams: the Pittsburgh Pirates and the Boston Americans. Barney Dreyfuss, owner of the Pirates, the winningest team in the National League, issued a "gentleman's challenge" to Henry Killilea, owner of the

Boston Americans, the winningest team in the American League, proposing a postseason championship, a World Series on a best-out-of-nine basis. The series would begin in Boston (three games), move to Pittsburgh's Exposition Park (four games), then back to Boston to complete the series. A deal was struck.

The Royal Rooters had arrived. The lives of these guys became The Game. The first World Series, played at Huntington Avenue Grounds, drew an overflow crowd of decorously dressed men, women, and children, mainly arriving in Union Station and on streetcars. (It was 1903, the same year Isabella Stewart Gardner, a baseball fanatic, opened her salon—as opposed to Nuf Ced's saloon —at Fenway Court.) It is open to question, and betting, whether the newspapers of the day spilled more ink on the Boston Americans or the Royal Rooters, not to mention who played the better game. McGreevey was a true showman, the stunts he and his guys staged an early form of performance art.

What ensued on the field—that day or almost any other—was often upstaged by the choreographed yet free-form pageantry of the Royal Rooters, all sporting flamboyant red silk badges. Mc-Greevey, a five-foot-tall dervish in a bowler hat, led the World Series march of the Royal Rooters to Huntington Avenue Grounds and to Exposition Park in Pittsburgh. He amused and infuriated, stepping out the Irish reel on the metal roof above the Americans' bench, exhorting his Rooters, the fans, and the Boston team. A popular song, "Tessie," became standard fare. The Rooters crooned the tune at every game, warming up the home team and their supporters, and exasperating the opposition. Their enthusiasm was so winning (and hilarious—these were grown men, chortling, campaigning, mugging) that even the fans on the other side often warmed to them. "Tessie" was hard to take, however; the Rooters changed the lyrics, to better skewer the Pirates.

In the end, the first World Series was no sweep, as in the come-from-behind Red Sox win of 2004. The Boston Americans lost the first game (played in Boston), won the second, lost the third (the

infamous "riot game," though the riot—essentially overly stimulated fans—happened early on), lost the fourth (played in Pittsburgh), won the fifth, sixth, seventh, and.... The day of the eighth
it was cold and gray. Rain threatened. The crowd was the smallest
of the Series. No matter—the Rooters arrived with a brassy, shiny
marching band, taking their special seats, as befits royalty.

The Bostons shut out the Pirates, winning 3-0. Boston pitcher
Bill Dinneen kept the Pirates from scoring a single run. The crowd
surged forward, tore onto the field, and engulfed the players, lift-
ing them onto their shoulders. The Rooters partied, drank, danced,
and sang. Third Base truly became home.

The official World Series was a keeper, not only because of the
interest it generated in the fans, all the new fans it made, and the
money, but because it legitimized the sport. The official World Series reformed baseball and set a new standard. Creating a truce and
treaty between the National and American Leagues, it established
a governing commission and a set of rules, including an agreement
to honor each other's players' contracts.

The Boston Americans became the Red Sox in 1907, named by
John I. Taylor of *The Boston Globe* newspaper family, and won the
Series in 1912, 1915, 1916, and 1918. Fenway Park was opened
in 1912. Nuf Ced McGreevey's Third Base saloon was closed by
Prohibition. The celebrity barkeep, who owned the building, then
leased it to the Boston Public Library (BPL); it became the Roxbury Crossing Reading Room. The relationship between Nuf Ced
and the BPL became much more than that of landlord and tenant;
McGreevey left his papers—including scrapbooks and scores of
photos that had been on the walls of Third Base—to the library.
You can view them today on microfilm.

The general outlook of Red Sox fans: Eighty-six years is a long
time to wait for a win. But not if you expect to lose as much as you
expect to win. You could definitely handle winning, but you get
used to losing. Still, you never lose so much that you stop thinking about winning; you never stop thinking about winning. You

used to win a lot. Your time will come again. And you aren't "not winning" because you aren't good. It's because something always happens, and life is unfair. If it were fair, you would win.

Every Red Sox fan could speak nonstop about his or her particular memories of beautiful moments and crushing disappointments—which were sometimes the same moment because of the way a baseball moves, with unpredictable velocity and uncertain trajectory. But victory had escaped the Sox fourscore and more years until a series of remarkable, basically impossible events. In 2004, the Red Sox became the first baseball team ever to win a seven-game postseason series after losing the first three games, one of the greatest comebacks in sports history. Boston swept the World Series, the 101st.

Everything was different from the first time—the ballpark, the salaries of the players, the prices of the seats, the ethnicities of the players and the fans, the electronic media, corporate sponsorship, endorsements, issues of performance-enhancing drugs, even some of the rules of the game.

And yet. The response of the fans—working men and women, and all the swells who'd joined the ranks (tens of thousands of them over the century, including Isabella Stewart Gardner)—was the same.

Victory is victory. Pride is pride. Joy is joy. Any guy at Nuf Ced's could have told you that. It was their specialty.

FIRST NIGHT

☜ 1 9 7 6 ☞

No offense to Boston (which anyone reading or writing this book dearly loves), but not so long ago, it was dull. Not only on New Year's Eve, but on the other 364 evenings.

Truth to tell, many of us loved the dullness, the soft secrecy of ancient urban corridors, crumbling brownstone buildings, parched ivy struggling up wrought-iron fences and around creaky gates. Well into the twentieth century, Boston was the kind of place where you did not apologize for not going out. A chair with a lamp was the home entertainment center. Yes, there were movies and libraries, museums and concert halls, and restaurants where you could order baked stuffed haddock served with tartar sauce. But if you wanted to have deep thoughts on New Year's Eve, or cut a rug, or have a visionary insight, Boston did not help. You would have to go to India for insight, New York City to boogie.

Which is not to say there were no cool people. There were artists, for one thing. Artists are always cool. But the artists were at home, or at little parties. And the city of Boston had been suffering all through the 1950s, 1960s, and 1970s—with the loss of jobs, dwindling real estate values, crime, racism, and the stigma of court-ordered busing.

But change was in the air. The sixties—for all the years of tumult—had energized Boston. The boundaries between professions and cultures were starting to blur, as young people were buying dilapidated row houses and restoring neglected neighborhoods.

Artists were moving into old factory buildings and warehouses. And in 1976, Boston was uncustomarily high on itself. Visitors had flocked to the city's Bicentennial and brought new pride to natives. The idea of updating Boston's heritage—consecrating and celebrating it—had made for unions among historians, promoters, and planners. A glamorous Parade of Tall Ships had taken place in Boston Harbor, and even the Queen of England had dropped by. Earth Day, which began in 1970, and the environmental movement were demonstrating the power and seditious fun of citizen involvement.

One enchanted evening in February 1976, several artists and a lone psychiatrist were eating dinner at Clara and Bill Wainwright's house. The Wainwrights, both artists, were live wires in the city. Clara Wainwright had already started to create and organize public arts events such as the annual Kite Festival, staged in Franklin Park—one of Frederick Law Olmsted's gems, though gone scruffy—a sprawling public space with borders in Dorchester, Roxbury, Jamaica Plain, and Mattapan.

Wainwright wanted something out-of-this-world for New Year's Eve. Something that would be more original and fun than getting drunk, and, for those of spiritual inclination, something that would offer inspiration for the new year. She disparaged watching Guy Lombardo and "the ball drop on TV," which she dryly described as "not a very novel way to celebrate." Why not troubadours and tambourines, jazz sets and classical interludes, mime, dance, gospel music, and the swirling Chinese Lion Dance? Why not cluster events in Boston churches, with Boston Common as the gathering place? "We wanted to use the city as a stage," said Wainwright.

The city of Boston gave full consent, but the plan was left mainly as a private, nonprofit affair. Organizers didn't push the city for support, and the city didn't hover. "The bureaucratic types didn't think it would amount to very much," remembers Wainwright. "People who'd crack down on you were not awake yet."

As December 1976 drew to a close, expectations for the nascent celebration were modest. There was slick, packed-down snow on Boston Common, along with single-digit temperatures and wind.

On December 31, 1976, sixty thousand people poured onto the Common and fanned out to performances at churches and halls. "Magical" was the word of the evening. The feeling that flooded the crowds was comparable to seeing a plain-Jane—or plain-Joe—office mate utterly transformed, newly glamorous, seductive, and appealing! The event was amazing for participants because of the array of artistic talent and the ease of moving from building to building—most of them old, welcoming, historic structures—and partaking of one artistic event after another. There were no tickets, no lines, no driving. (In the beginning, all events were free. Later, one bought a button and was admitted everywhere.) Revelers walked from place to place and greeted other revelers. Most of the performance spaces were small, making for intimate gatherings. Artists, especially visual artists who generally worked alone, were heartened by the quantities of inquisitive, enthusiastic, and appreciative people. Performers were thrilled by the successive waves of audiences who showed up for sequenced performances—like playing to four packed houses per night. The crowds, which included many individuals who had never before attended a live performance, were dazzled, and generous with applause.

It was a totally original event—born on a dining room table—and from the start a phenomenal success. Wainwright and other volunteers had dreamed up, created, and organized an annual, outdoor, wintertime urban arts festival, a smorgasbord of performing arts activities, exhibitions, and site-specific creations (including lavish, large-scale ice sculptures). First Night became a tradition and spurred similar celebrations in hundreds of North American cities and towns. It helped to bring people into Boston, and created city boosters and arts patrons. For decades, in the depths of winter, it has entertained, sustained, and inspired.

Especially in the early years, there was a feeling of pageantry and progression as one went from event to event, taking in the eclectic offerings and making the metaphoric leap to the possibilities in store for the new year. Even the intervals and the choreography of attending events became art events! First Nighters would be outside, in the dark, huddling in the winter cold under streetlights consulting the printed schedules, deciding which event to go to next (scores of events happened simultaneously at different venues). Then suddenly they would be in a bright, warm, inviting setting—Trinity Church with its stained-glass windows, or the Victorian mansion of the Boston Center for Adult Education, or the holly-bedecked parlor of the French Library—taking in the grandeur of the building and the high spirits of the audience. Then, an entrancing live performance, often at close range—anything from bell ringing to barbershop harmony to the reading of holiday haiku. Then back to the cold streets and bare elm trees on the Commonwealth Avenue Mall, consulting one's schedule once again —over and over, till midnight, when lavish fireworks were displayed over the Charles River, summoning in the New Year. Clara Wainwright's idea—of being in the city with other people and having fun, while connecting with something timeless and sacred—became First Night.

Inevitably, as the event became an institution, its character changed. It is no longer intimate and outré, there are lines for popular events, and a button costs fifteen dollars. The crowd of sixty thousand has swelled to over a million; the budget in 2004 was 1.2 million. The magic of the procession is still there, and the fireworks display, and the array of local artists and performers. But, especially since the terrorist attacks of 9/11, security has become an issue, as it must now be whenever crowds gather in American cities. Raising funds to pay for police protection has become a stumbling block. Many of the outrageous, free-form events that used to happen no longer do.

Clara Wainwright, godmother and former executive director of First Night, has assumed a kind of forever-emerita position. Still a freethinker, she has speculated on the idea of a fringe First Night, to summon back the serendipity of the early years. And who knows, there could be a dinner party of artists and a lone psychiatrist sharing good food across a table—perhaps the original table, a lucky table—sparking another millennium's dazzle and enchantment.

INNOVATION
& ADAPTATION

ICE

Simple things, everyday things—how we take them for granted. We turn on the tap; there is water. We flick on a switch; there is light. Tropical juice is frozen in cans. Chocolate ice cream is in a box. Cubes of ice stay neatly in their trays in our refrigerators (the refrigerator may even create them), ready to chill drinks, relieve pain, or reduce fever.

Once there was no ice, even in the most civilized countries, unless they were cold countries and it was wintertime. Or, if you were a king in a mountainous country, and desired cold desserts and drinks in summer, you could send ice couriers into the mountains to bring chunks down the perilous mountainsides for your dining pleasure. Some couriers would tumble into ravines, of course, along with their pack animals, but still, you as monarch would have cool scented drinks and refreshing desserts.

In early New England, farmers with ponds would dig up thick chunks of ice and pack them in layers of sawdust and straw for storage in cellars and barns, keeping the ice well into summer. Fresh Pond in Cambridge was a prime source of ice, as was Jamaica Pond in Jamaica Plain, Boston. (Some winters, Fresh Pond would freeze three times, providing three lucrative crops!) Well into the twentieth century, even in developed countries such as the U.S., ice was delivered to people's homes to be placed in iceboxes, our pre-electric refrigerators. Many of us still have a relative who refers to the refrigerator as the icebox.

Many great ideas—scientific, artistic, mercantile—come from someone noticing and taking advantage of natural phenomena and distributions. Frederic and William Tudor, two Boston brothers, born not long after the Revolutionary War, noticed that Boston and its environs had lots of ice, a surplus of ice, one might say. But the Caribbean, where people might have enjoyed ice cream, had none. The southern U.S., where ice seemed a natural for mint juleps, had none. India, where the British and Indians presented a huge potential market, had almost none. These perspiring multitudes not only lacked ice, they didn't know what they were missing.

It was the older Tudor brother, William, a bookish Harvard graduate, who came up with the outrageous idea: to transport ice on ships. But it was the younger brother, Frederic (1783–1864), who had the moxie. The idea of shipping ice seemed absurd to most traders of the era. They mocked Frederic Tudor. But he would not give up. The young Tudor—small, stubborn, self-styled, energetic—not only risked ridicule, he took it on wholesale in his conviction that ice could be packed, preserved, and shipped, and that he could create a market.

If Tudor had described his concept and plan in the abstract, as in a Harvard Business School case, he would have been applauded. It was at once brilliantly simple and innovative. Ice was a fine thing, a grand thing, enabling you to store food, cool drinks, and treat fevers and flu. It cost nothing to grow, and bodies of water were not owned in Massachusetts; they were public property, their contents up for grabs. In addition, ships routinely showed up with cargo in Boston and left with nothing but ballast. Boston produced no crops worth taking back to their ports of origin. Instead of loading up with ballast—a worthless cargo of stones, dredged up from Massachusetts Bay—why not ice, Frederic reasoned, especially as most of the ships were coming from hot climates whose populations could be converted to appreciative consumers of ice.

Tudor sent his first ship to Martinique in 1806 in the care of his brother William and cousin James. Scion of a wealthy family, though one that would come to ruin, Frederic was confident, even cocky, his outlook derived from economic advantage, temperament, and propitious times. He was so nervy that for the voyage to Martinique he bought a brig named *Favorite*, using up most of his capital. This was never done; one bought space on a ship and paid for freight. But shipowners were reluctant to carry "frozen water," which could damage other cargo as it melted, and Tudor felt he needed to outfit a vessel, to insulate the hold, to minimize meltdown. Financially, this initial venture was a disaster. But Frederic felt he had learned from the experience—the bold, cold transport and sale of *glace*, a twenty-one-day voyage with 130 tons of ice—and would do fine in the future.

Throughout his long life, Tudor drew on family connections and those of extended family. When family money was available, he drew on it. When that fortune was extinguished, he traded on the fortunes of friends, his attitude that of a trader/merchant/banker: capital was resource and product, as well as reward. As soon as money was made, it was reinvested, like planting more trees in an orchard when the fall harvest does well.

While there were numerous technical problems to be solved—including the building of aboveground icehouses for commercial storage, educating buyers on point-of-purchase storage, and using horse-drawn saws to score ponds for efficient cutting—the most significant development was the idea. Before Tudor's operation, it was not only tropical countries that lacked ice, it was the southern part of the United States.

The course was uneven, one of continual mishap and struggle, but Tudor persevered, and eventually developed reliable markets in Charleston, Savannah, New Orleans, and Havana. Had he failed, which he almost did many times, he would have been labeled a crank. As he kept going, he is considered a maverick genius. He amassed a fortune and became known as the Ice King.

The story of the ice trade is a marvelous one, made for a keen observer who can weave a story with multiple threads. Gavin Weightman, a British journalist and filmmaker, authored *The Frozen-Water Trade* (2003), which tells the story well. ("Frozen-water trade" was the down-to-earth term used by seafaring men for the shipment of the slippery cold cargo.)

Weightman spins the yarn of Tudor's difficulties and eventual success, presenting him as a classic, canny New England merchant, intent on making something out of nothing. It took him decades to do it, however, and he suffered not only ridicule and insolvency, but illness, including a mental breakdown in 1821. Many a time Tudor was just one step ahead of the sheriff and debtor's prison, and sometimes insufficiently fleet, forced to suffer added humiliation and impediments to his grand plan. (He noted in a diary entry around 1834 that he had lost four times what his wealthy grandfather had been worth.) He would be on the edge of making a killing —with a cargo of melting ice—and be leveled by a trade embargo because of one war or another. He would line up his ducks, and there would be a warm winter, with insufficient quantities of ice or depleted summer stores. In August 1819, he was so exasperated— but undaunted—that he sent a frigate to Labrador to harvest an iceberg. Miserable seamen tracked ice floes, then clambered aboard icebergs, hacked off chunks, and loaded them into small boats in the frigid sea. An iceberg fell on the brig, almost sinking it. But weeks later, Captain Hadlock and the brig *Retrieve* out of Castine, Maine, landed in Martinique so that chips of an iceberg might make dessert.

Weightman's book carries a quote from one of Tudor's diary entries, written in 1817 and later inscribed on the cover of his ice-house diary:

"He who gives back at first repulse and without striking the second blow, despairs of success, has never been, is not, and never will be, a hero in love, war or business."

As Tudor became successful, marrying for the first time in his fifties, he continued to find opportunities to make something of nothing. He applied his curiosity, energy, and outrageousness to more than ice: salt, for example, coffee, and lead. He started a salt-works near his summer cottage on the Massachusetts island of Nahant. The wooden icehouse he created for use in Havana, which used sawdust and peat as insulation, became a prototype. He got into the manufacture of home iceboxes, as well, and could not resist tucking into his cold, southern-bound cargo the Baldwin apples he grew at his home in Nahant, along with walnuts, grapes, blocks of butter, and cheese. His longtime colleague, Nathaniel Wyeth, invented the ice plow, employing horses, which revolutionized the harvesting of ice.

The apotheosis of the Ice King came about in 1833. Tudor successfully shipped ice to India—over a hundred tons of glistening chunks that had traveled sixteen thousand miles and taken four months to arrive. A great hullabaloo took place in the harbor of Calcutta. The near magical, coveted, long-awaited cargo was accorded every consideration—allowed entry duty-free and unloaded at night (the day temperature being around ninety degrees). The agent, Mr. William Rogers, was presented with a silver cup by the British high-mucky-muck, Lord William Bentinck, Governor-General and Commander-in-Chief of India. The cup's inscription, a graceful tribute, includes the words, "in Acknowledgement of the Spirit and enterprise which projected and successfully executed the first attempt to import a cargo of American ice into Calcutta—Nov 22nd 1833."

Following this theatrical success, and with effective advertising and marketing, Tudor became a fabulously wealthy man, an ice magnate. He pioneered and built a major business, an industry with its own expertise and technology. The frozen-water trade employed thousands and shaped American recreational tastes—from mint juleps and sherry cobblers to the corner spa and ice-cream socials. It changed economic and cultural history in overseas colonies. Ice became a luxurious necessity, a necessary luxury. And not any ice, mind you, but New England ice, considered the finest. Originating in spring-fed ponds, so clean and fresh it could be placed directly into drinks. From Jamaica Pond to lemon squash.

BOSTON CREAM PIE

❦ 1 8 5 6 ❦

It looks like a cake and has the name of a pie. It was invented and introduced at Boston's snazzy Parker House Hotel in 1856. There's a fancy-pants hotel version and a bake-at-home counterpart so popular that Betty Crocker grabbed it in the 1950s and turned it into a supermarket cake mix. Regarded from a scholarly perspective, the venerable cake-pie provides culinary history and insights into agriculture, advertising, and Boston character. Regarded as dessert—a freshly cut, vanilla-scented wedge on a plate —the cake-pie is irresistible, a mix of elegant and homey, deluxe with a cup of dark roast coffee, delicious with a glass of milk.

Boston cream pie (BCP) is a two-layer butter cake with custard or cream in the middle and sleek chocolate icing on top. The sides are often kept naked, icing-free, the better to tempt you with a view of the custard in the middle, a kind of peekaboo blouse for cake. At the Parker House, they slick the sides with custard and dabble on toasted almonds, very pretty, and what they've been doing for 150 years. If you buy BCP in a bakery, where it is often displayed on a glass pedestal, the layers will be a little higher, and a pastry chef may have streaked decorative white spirals or a delicate web pattern over the chocolate icing to Frenchify the cake, which is true to its origins.

This dapper dessert even looks Boston-style, that know-it-when-you-see-it merge of sexy and preppy, a neat little cake with a shiny top and sensuous pastry cream layer.

Here's how it came about: In the nineteenth century, the French were baking away, and American cooks were, too. BCP has its culinary roots in both cultures. The Parker House dessert, introduced by a French chef at the hotel, used a butter sponge cake, pastry cream, the aforementioned almonds pressed into the side of the cake, and a chocolate fondant icing. (It was actually the icing that created the sensation. At the time, chocolate was considered a drink or used in pudding, rarely in baking.)

Dessert lovers adored the cake and clamored for it at home. But it was a pain to make. The Parker House version required fancy ingredients and French cooking skills. Fondant, the icing, had to be first cooked as a candy form, then cooled, then kneaded, then melted for icing, at which point it was likely to harden too quickly and cause home cooks to scream and drink the sherry.

Meanwhile, across America, a dessert called Washington pie was being concocted, served, lapped up. This simple two-layer cake, which appeared in early nineteenth-century American cookbooks, had jam or preserves between the layers and was sometimes dusted with confectioners' sugar.

Enter Boston ingenuity, truth, justice, and the American way. In an admirable article in *Gastronomica*, writer Greg Patent follows the culinary evolution of BCP and cites it as an example of democracy. "Americans have long sought to emulate chefs," he says. "We feel we have the right to create exceptional food in our own, often unexceptional kitchens.... All we need are specific instructions and some basic equipment, and we can handle the rest."

Undeterred by French culinary techniques, Boston cooks and others adapted the recipe for Washington pie, using its white or golden layers, substituting custard or cream for jam between them, and decorating the cake with confectioners' sugar or a homestyle version of chocolate icing—never mind the folderol of that annoying chocolate fondant.

If you look up the recipe in a *Fannie Farmer Cookbook*, you'll be advised to use a basic butter-cake recipe (one is called Boston Favorite Cake), to then place custard filling between the layers, and to spread the top with chocolate frosting, or to sprinkle it with confectioners' sugar.

The esteemed cake has made for good politics. (Would a cake be truly Bostonian if not put to political use?) In 1996, Boston cream pie became Massachusetts's official state dessert—in spite of strong opposition from the tollhouse cookie lobby. Norton High School students drafted the bill as a history class project and spoke ringingly of the virtues of BCP, though others, perhaps small-minded, championed the diminutive cookie. Representative Joseph Sullivan (D-Braintree), a friend to all desserts, was asked during the controversy which he would eat, if offered both. "I'd have the chocolate chip cookie on top of the pie," said Sullivan.

The ceremony to sign the pie into law was held at the Omni Parker House, a few blocks from the State House. Governor William Weld, a tall man, a Republican, did justice to the hefty slice he was served. "Boston Cream Pie is one of my earliest memories," said Weld, savoring the memory. "It's thanks to Boston Cream Pies that I am the biggest politician on Beacon Hill and getting bigger all the time."

AUTHOR'S NOTE

You can make a facsimile of the fancy-pants hotel version by doing what nineteenth-century home cooks did—improvise. Bake two eight-inch or nine-inch butter-cake layers, adhere the layers with a simple boiled custard—made in a double boiler—and lap chocolate glaze (not thick frosting) over the top, letting it drizzle down the sides. Very nice with a glass of cold milk, a mug of hot tea, or a goblet of sparkling wine.

BOSTON TERRIER

If the port of Boston were graced with a signature statue, one espousing political principles and ideals, our statue of liberty might feature a small, brave, intrepid dog. The dog would be the sturdy, tidy, highly intelligent, altogether *chahming* Boston terrier, the first dog bred in America, a native dog-son that appears to be wearing a smooth fur tuxedo and that proved too smart, friendly, and personable to engage in the fighting sport.

He descends from English bulldogs and terriers. His ancestors were likely "borrowed" by Beacon Hill servants who had grown fond of their employers' pets and saw possibilities for interesting genetic combinations in the pugnacious character of the bulldogs and determined scrappiness of the terriers.

What came of these well-intended but casual crosses was a kind of cosmic joke. The early breeders had in mind a dog that would fight. The affable breed that evolved was remarkably intelligent, highly sociable, and determinedly pacifistic. It is irresistible to note that the dog was a pol in the best sense—fun loving, gregarious, good-natured, with an instinct for compromise, and also handsome, clean, and mannerly, a nice chap to have about. He became known as the "Boston gentleman." While his bulldog ancestors were still snarling and wheezing, and the terriers in his family tree nipping and ratting, the gent went on to become a cherished companion, favored not only in Boston, but across the U.S. and in Canada and Europe.

The breed's early history provided ample staging for the best and worst behavior of his human companions. Many breeders scoffed at the newly introduced Boston terriers, insinuating that the dogs and their breeders were from the wrong side of the tracks, and refusing to acknowledge either. In fact, the breed *was* born a workingman's dog—adding to his allure in many quarters. He came into genetic existence in the taverns and stables at the bottom of Beacon Hill, where working people lived in the 1860s and 1870s. In that era, pedigreed dogs originated in Europe and Asia.

Initially, the dog was snubbed by the American Kennel Club. But his innate charm and the tolerant camaraderie of "dog men"— early twentieth-century dog annals allude to the sociability of "dog men" and "dog women"—sparked friendships among dog lovers of varied ethnicities. The dog they all found irresistible was small and friendly with a smooth black-and-white coat, sometimes brindled; a strong, compact body, with thick neck and short tail; pointy erect ears; and a square skull. He became known as "the Boston," and was stocky but dapper, a gent.

For decades, established dogdom refused to accord him the status conferred upon earlier breeds. Members of the Bulldog Club of America were especially antagonistic—pugnacious, even pugilistic.

The late nineteenth century was "clubby," protective of terrain; a slew of American dogdom customs and institutions were evolving. In 1884 the American Kennel Club was established, and in 1891 the Boston Terrier Club. Finally, in 1893, "the Boston" was admitted to the American Kennel Club, the first American-originated breed to receive kennel club recognition.

E. J. Rousuck wrote an entire book on *The Boston Terrier*, published in 1926. Though his racist references are grating to modern ears, the author describes his growing attraction to the terrier, in part based on observations of the dog fanciers the breed attracted.

Rousuck attended a democratizing dog show at Horticultural Hall (across from Symphony Hall) and reported his own change of heart in a chapter called "The Melting Pot of Dogdom."

> The great significance of what appeared before me in this small, roped off arena, became the nucleus of an ever increasing respect for the dog there represented: a dog which could bring into a single space of just a few feet across, a Jew, an Irishman, a negro, a German, an Italian and a Chinaman.... it seemed incredible that all these different types and colors should meet together inspired by a single interest, the Boston Terrier.
>
> A dog as democratic as his mother-land herself. Dog of window washer, pal of laundryman, idol of priest and laborer, Jew and Gentile. Literally did livery stable and Bar Harbor meet, all differences forgotten.

The annals of Boston terrier history overflow with photos of endearing dogs, but perhaps none so memorable as Champion Lady Dainty, born July 9, 1902 (AKC no. 71,617). Lady Dainty weighed just fifteen pounds, was white all over with dark brindle markings on her head, a neat white blaze, and a tight screw tail. Her liquid gaze melted hearts.

In 1975, Governor Michael Dukakis signed a proclamation making the Boston terrier the Bicentennial Dog of Massachusetts. Speaker Tip O'Neill quickly followed suit, shepherding legislation through the Congress to make the dog the Bicentennial Dog of the United States. In 1979, Governor King was persuaded to sign a bill making the Boston the Massachusetts State Dog. Governor King was a pushover. He had grown up with a Boston named Skippy.

It is difficult to convey the combined hilarity and dignity of a large group of Bostons and their owners, their staunch democratic appeal. But you know it when you see it. In April 1991, a most festive event, an energetic sea of black and white, was held on Cape Cod to recognize the hundredth anniversary of the Boston Terrier Club of America. In the lot of the Tara Hyannis were shiny, expensive foreign cars, battered station wagons, spiffy antique roadsters, Jeeps, trucks, and vans, arriving with their passengers, their gents—eager, personable, and sweet, born in the USA.

FANNIE FARMER'S *BOSTON COOKING-SCHOOL COOKBOOK*

❦ 1 8 9 6 ❧

Fannie Merritt Farmer (1857–1915) was the Julia Child of her day. She loved to eat, to educate, to fuss over food. With enthusiasm, charm, and élan, she communicated her ardor to others and taught the way to cook—to understand ingredients, equipment, tools, and techniques, and to prepare delectable food at home. She was personable and engaging, but also scientific and practical.

Fannie became a national craze, changing the way America cooked, ate, and entertained, and even how they thought about cookery and food. She had women all over America carving lemons into tiny baskets—how else to serve Harvard Salad?—and serving dainty appetizers flanked by saltines, standing up like barricades.

She was a hometown girl, born and raised in Boston. A physical handicap probably determined her future. At age sixteen, her left leg became paralyzed—most likely because of polio—and in the pre–disability rights world of the 1870s, she was unable to continue her education beyond high school. Smart, game, and needing to help support her family, she went to work as a mother's helper in the home of a family friend. She became intrigued by cooking—not only by the tastes, but by the way the ingredients combined: the alchemy of cooking.

Red-haired, blue-eyed Fannie became a celebrity teacher at the Boston Cooking School, in the vanguard of American cooking schools, established by the Woman's Education Association in 1879. In a time when women were still sheltered, she traveled around the country giving lectures and workshops—attended by middle-class women in department stores, and by nurses, dieticians, and physicians in hospitals.

Her best known book, first published as a textbook, the *Boston Cooking-School Cookbook* (1896), became known as the *Fannie Farmer Cookbook*. When she died in 1915, over 360,000 copies of the book had been sold, and it was still being reprinted in runs of 50,000. In addition, she wrote *Chafing Dish Possibilities* (1898), *Food and Cookery for the Sick and Convalescent* (1904)—which she considered her most important book—along with *What to Have for Dinner* (1905), *Catering for Special Occasions, with Menus and Recipes* (1911), and *A New Book of Cookery* (1912).

She was a prominent magazine columnist, a personality who endorsed brands and culinary products. Like Julia Child, she was not a gadfly, but a serious woman with a mission—who liked fun. She was an exemplar of scientific cookery, the nineteenth-century movement to elevate and standardize culinary practices and to teach them to immigrants and the poor. But it didn't stop her from loving dessert.

Her good sense continues. Find a copy of the *Fannie Farmer Cookbook* (any incarnation will do); commence cooking. Speaking across the centuries, Fannie starts you off with the basics—ingredients, techniques, cookware—and then gets into the recipes, thousands of them, set in reassuring bookface type, with nary a glossy photograph, but with charming drawings of cookware and tools ("A Few Helpful Objects") and the occasional favorite foodstuff, such as Icebox Cake. Oftentimes there are bits of advice (though perhaps not as homey and encouraging as those of her predecessor at the Boston Cooking School, Mary Lincoln).

She is still a household presence, a hovering kitchen goddess. In many American families, she is known as "Fannie," as though she were everyone's aunt. Her book and recipes are also "Fannie." And so, when the source of "egg in a hat" is questioned, the cook will say, "Fannie," in much the way that, when asked about the source of coq au vin, the cook will say, "Julia." Our timeworn recipes for New England comfort foods—Boston cream pie, Indian pudding, Boston brown bread, even macaroni and cheese—are all thoroughly "Fannie."

While Fannie's primary mission was teaching housewives how to prepare delicious, nutritious food at home, and standardizing American cooking, she also focused on teaching society gals how to prepare fashionable menus, and worked with nurses and dieticians to develop menus for convalescents. She marketed her talents and skills, "branding" herself as a culinary authority able to offer expertise in both popular and specialized settings.

Her supposed claim to fame is that she was the first to introduce a scientific approach to cookery. This is not really so, though her hold on the title is secure. She built a great deal on the work of Mary Lincoln, whose own textbook for the Boston Cooking School preceded Fannie's by twelve years. *Mrs. Lincoln's Boston Cookbook* (1884) was more original, even radical, than Fannie's but for a variety of reasons did not seize the popular imagination. While Fannie's text used a more concise style, it may also have been that Mary Lincoln was ahead of her time, that she paved the way for Fannie.

Mary Johnson Bailey Lincoln (1844–1921), born in Attleboro, Massachusetts, daughter of a minister, graduated from Wheaton Female Seminary (later called Wheaton College) and went on to study and teach at the Boston Cooking School. Her book, a text for the school that was also marketed as a cookbook, was highly original—organized, comprehensive, illustrated. It had personality and a mission.

Describing her selection of recipes, Lincoln wrote, "They [recipes] must include the most healthful foods for those who have been made ill by improper food; the cheapest as well as the most nutritious, for the laboring class; the richest and most elaborately prepared, for those who can afford them physically as well as pecuniarily."

She became a regular on the national lecture circuit, a culinary celebrity before Fannie. Hundreds of women gathered for talks at cooking schools, colleges, women's clubs, department stores. She co-founded the *New England Kitchen Magazine*, was a lively participant in the New England Women's Press Association, and had a strong entrepreneurial spirit. She endorsed brand names of popular products, including a household staple that bore her name: Mrs. Lincoln's Baking Powder.

Before the innovations of Lincoln and Farmer, measurements were wildly inexact, involving handfuls, spoonfuls, cupfuls ("cup" being anything from a teacup to a shaving mug, filled, partially filled, leveled, or overflowing). Terms such as "butter the size of an egg" or "heaping tablespoon-full" were common descriptions of quantity. Recipes were written in run-on prose and barely punctuated. The nice neat ordering we take for granted—a list of measured ingredients followed by a list of procedures—is a combination of Fannie Merritt Farmer, based on the work of Mary Johnson Bailey Lincoln, and a contemporary, Mrs. Sarah Tyson Rorer of Philadelphia.

Those miracles of measurement, metal half-pint cups with gradations, and measuring spoons, made their way to market during the 1880s, around the time Fannie began her training at the Boston Cooking School. As Laura Shapiro points out in *Perfection Salad: Women and Cooking at the Turn of the Century*, Fannie used these tools from the start of her professional training. It seems like simple stuff to us, but measuring tools were a godsend to home cooks struggling to cope with imponderables ranging from woodstoves with uncertain and fluctuating temperatures to ingredients of varying quality—this was before widespread standardization—to poverty and unhygienic conditions.

Though often attributed to Fannie, standardized measurements were part and parcel of Mrs. Lincoln's earlier book. Even the practice of leveling (topping-off with a knife) and the use of measuring cups and spoons were employed by Mary Lincoln, if inconsistently. Which is not to say that Fannie did not have her own "eureka" moment concerning reliable measurements. When she was in her twenties, working as a mother's helper, the little girl she was caring for asked what "handful" and "pinches" and "heaping teaspoonful" meant. In explaining, Fannie realized how much better it would be if these approximations were exact. She took her insights and Lincoln's groundwork and ran.

Similarly, a century later, Julia Child was not the first French chef to appear on television. A combination of factors—her personality, her approach to the material, and the public's growing interest in cookery—produced the successful chemical reaction. Julia's exuberance was irresistible. In her own way, she was sexy, as was Fannie, once her love of food came through.

Julia Child and Fannie Farmer both enjoyed rich foods. Mary Lincoln preached "hygienic living" and looked askance at sauces, condiments, frilly desserts, and presentation folderol. By contrast, as reported by Laura Shapiro in *Perfection Salad*, Fanny Farmer took Lincoln's nourishing but plain-Jane fish fillets and lapped them in cream sauce. She inserted Harvard Salad (chopped chicken with whipped-cream dressing) into faux baskets carved from lemons. She promoted chocolate pudding, chocolate sauce, and caramel bisque ice cream. It is easy to imagine Julia and Fannie dining together, and tempting to imagine their menu.

Viewed through the lens of social history, cooking in America during the late nineteenth century was much more than cooking. (*A Thousand Years Over a Hot Stove: A History of American Women Told Through Food, Recipes, and Remembrances*, by Laura Schenone, is a fascinating book.) Women were tiptoeing into equality, a kind of genteel, ladylike warfare, consisting of discussions and forums, meetings and marches, regional congresses and confabs among mainly middle-class women, with occasional fusillades that dropped all pretense of polite acceptance of the status quo. Education for women, volunteer work, labor legislation, child welfare, abolition, and above all, suffrage, were avidly discussed by progressive, Victorian-era women. Improvements in science and technology, including the railroad, telegraph and telephone, electric iron, and light bulb changed the way everyone thought. These inventions stimulated scientific cooking, a domestic movement that women of varying opinions could coalesce around. It was not so much that Mary Lincoln and Fannie Farmer were trying to foment revolution, though they were certainly in the vanguard; it was that the currents of science and the collective vigor of American women had affected their thinking, inspired them, and characterized their work.

Fannie lived to be fifty-seven years old. Her ashes are buried in Mount Auburn Cemetery in Cambridge. If you buy an "old" *Fannie Farmer*, one published before 1965, you'll get a more authentic Fannie, a direct descendent of the original, likely to have recipes for Egg with a Hat (not to mention Stuffed Eggs in a Nest) and Cheese Toast (or Milk Toast I, Milk Toast II, Brown Bread Milk Toast). Cream Toast can be found in the original 1896 book. But even Marion Cunningham's modernized version (1979) carries old-fashioned New England cookery, including a fish chowder that uses fish heads and frames for stock, cubes of salt pork for flavor, cream for sensual delight. One can almost hear Julia exclaiming over Fannie's dish—the rich cream, the white chunks and slivers of flaky cod—and hear her throaty lilt enunciating, declaiming, and celebrating the fish frame. They would have made a great team. In a way, they do.

DUDLEY STREET NEIGHBORHOOD INITIATIVE

❦ 1 9 8 8 ❧

The neighbors of Dudley—African American, Latino, Cape Verdean, and white; men and women; old and young—did more than "take back the street." They made a new street, a better street, with tidy houses, pocket gardens, and parks, two new schools, two new community centers, businesses, and a town common with banks of beach roses where there had once been heaps of garbage and broken glass. They did it by gaining control of hundreds of land parcels disfigured by arson and illegal dumping—vacant lots that had been trashed and abandoned. Thousands of neighbors banded together, took control of the land, and created an urban village.

In 1988, the Dudley Street Neighborhood Initiative (DSNI) of Boston became the first nonprofit community organization in the nation to be granted eminent domain over private land.

The story of DSNI is a story about a law: an innovative interpretation and application of eminent domain, which ordinary citizens used to acquire land to build housing for poor people. The story is about people—the Dudley neighbors and their years of hard work, organizing their community—and the human capacity to restore dignity and hope, reclaim damaged land, solve problems collectively, and recover a neighborhood. The story is most visibly and dramatically about 440 new units of housing—many of them in attractive woodframe row houses—along Dudley Street and else-where, and 560 rehabbed units in older buildings.

"The legal structure that allowed us to do this is not our real power," said John Barros, executive director of DSNI since 1999. "Our real power base is the influence we have as representatives of the collective opinion of a community, and that is what allows us to move on behalf of the public good."

The land, law, and the Dudley neighbors came together in the Roxbury/North Dorchester area of Boston, one of the city's poor-est neighborhoods, during the 1980s and 1990s. Dudley Street— a long, old street, named for Puritan governors—meanders east from Dudley Square in Roxbury, the hub of Boston's African Amer-ican community, into Uphams Corner in Dorchester. The well-trafficked street, just a few miles from downtown Boston, was blighted by hundreds of filthy, dangerous abandoned lots. It had been that way for decades, ugliness all around, and was worsening by the day. Landlords neglected and abandoned rental housing. Buildings were routinely torched. Cars and dead refrigerators and stoves were thrown into the filthy lots. Crime and drugs were ram-pant. "Arson for profit, red-lining, disinvestment," Ros Everdell, DSNI organizing director since 1988, clicks off the reasons for Dudley's decline.

Slowly, painstakingly, community activists pulled together a group of residents to try to improve matters. (Pulled together is not merely a figure of speech; leaders would sometimes go house to house, pulling neighbors to meetings.) In 1984, the Boston-based Riley Foundation paid a visit to the community and was so alarmed by its degradation—trashed lots everywhere, like a body infected with and disfigured by disease—that they immediately committed funds to improve conditions. They prepared a plan and invited comment. At the now legendary meeting, the people of the neigh-borhood—burned by past offers of help from outsiders—demanded to run the show themselves. The Riley Foundation agreed.

In 1984, the Dudley Street Neighborhood Initiative was formed to rebuild a core area of Roxbury and North Dorchester that has Dudley Street as its spine. At the time, about 24,000 people lived in the 1.5-square-mile area, about 33 percent of residents under the poverty line, half of the households headed by women. A board was elected with representatives from local agencies, businesses, religious groups, and residents from all the community's ethnic groups. Ché Madyun, a mother of three, was DSNI's first president.

At issue was not only the squalor of the neighborhood, its abandonment by property owners, and the social problems of many residents—poverty, drug addiction, lack of education and job skills—but the complicated ownership of the wrecked pieces of land. There were hundreds of vacant buildings, boarded-up storefronts, and trash-filled empty lots owned by individuals as well as the city of Boston. Ros Everdell remembers "entire streets closed-down by Jersey barriers" because they had become so degraded. DSNI, gaining credibility through their early accomplishments—putting an end to illegal dumping, towing away hundreds of abandoned cars, getting new street signs and traffic lights installed—was able to convince the city of Boston and the Boston Redevelopment Authority to grant them the power of eminent domain over hundreds of private and city-owned parcels, 1,300 parcels in all.

Eminent domain is traditionally the power of government to take private property for public use—typically a public works project such as a road or a dam—with fair compensation awarded to property owners. In the case of DSNI, a consortium of individuals gained control of (mainly) privately owned land. In the years that followed, the Ford Foundation provided a $2 million low-interest loan for the grass-roots group to buy land from private owners. The city of Boston kicked in fifteen acres of additional intermingled city-owned properties, a substantial allotment in this small, tight-knit urban neighborhood, and a good-faith gesture that facilitated planning.

Savvy DSNI members—who had become virtual real estate developers for the poor—recognized that their renewal projects would escalate in value and be "stolen" by outsiders, denying neighbors the living space they had fought for years to develop, because they could not afford the market rates. DSNI came up with the idea of a land trust, called Dudley Neighbors. A ninety-nine-year "ground lease" literally underlies many of the new housing units. These properties are owned in common. Resale prices of homes within the trust are restricted; housing stays affordable.

In the spring of 1996, Dudley Town Common—the gateway to the revitalized Urban Village, according to its prominent plaque —was dedicated. The Common is actually two parks, smack-dab in the community's center, one at the corner of Dudley Street and Blue Hill Avenue, with a bandstand, and the other at the corner of Dudley and Hampden Streets across from red-brick Saint Patrick's Church. (The church flies a "Welcome" banner, and offers three Masses on Sundays: Cape Verdean at 9:30 a.m., English at 10:30 a.m., Spanish at noon.) The sloped edge of the park with the plaque is landscaped with fragrant pink roses like those that grow wild on the beaches of Cape Cod. Once, these lots were piled with stinking garbage. The gateways and streetlights of these two welcoming parks echo the Victorian ambiance of restored Dudley Station, an MBTA transportation hub a few blocks away.

The neighborhood is now dotted with spiffy wood-frame row houses in shades of peach, sage green, and colonial blue. Several small brick apartment buildings and triple-deckers have been renovated. An old brick mill building has been restored and contains the offices of DSNI.

Dudley is very much a work in progress. Businesses are sparse. Some houses are still boarded up. The accomplishments of two decades may be difficult for an outsider to grasp without having walked these streets over time. Even during the worst years, there were some decent houses and people coping as best they could. But the overall impression was one of neglect, desolation, and decay, and also stark contrast: well-scrubbed children on their way to school walking past overturned car carcasses, jagged broken glass, and trashed refrigerators with their rusty doors still hinged. Other lots were littered with syringes, condoms, and flaky, painted clapboards stripped from houses—lead paint chips leaching into the soil. A passerby could see dead rodents that had eaten poison, then skittered from buildings in search of water. These sights are largely gone now, though vacant lots remain. Across from Dudley Common, behind St. Patrick's, a sizable field with waist-high, waving wildflowers looks like something from earlier, pastoral Roxbury. Ros Everdell looks over the Queen Anne's lace, chicory with bright blue flowers, and milkweed, and sees housing. "The property belongs to the church [the Boston Archdiocese], and we are working on the idea of housing for the elderly...so people who grew up here could stay."

Like a quilt, the work in the neighborhood happens piece by piece. There is an overall design, and ongoing community meetings (at St. Patrick's), but because of limited funding, there are continual improvisations. As "material" becomes available, new housing is constructed, old housing repaired, and then the pieces are sewn into the community. The residents of these new and restored houses—often longtime local residents—cherish their homes, their place.

In the spring of 2005, a longtime dream—a ten-thousand-square-foot commercial greenhouse—came into fruition, built on the site of a former auto body shop. How fitting—as Dudley is a community of good cooks—that the enterprise grows green garlic, the pungent herb much in demand by restaurant chefs. The epicurean ingredient is to be distributed by a Rhode Island garlic producer. Down the line, the greenhouse will raise salad ingredients to supply Dudley neighbors.

The Puritans of seventeenth-century Boston would not recognize many of the people of today's Dudley—African Americans, Cape Verdeans, Latinos. They might not recognize green garlic, either. But they would recognize enterprise, ingenuity, community, and grit. They would recognize a compact to self-govern and a town common, especially one sited near a church. (St. Patrick's, built in 1906, has a Star of David in the window; it was originally Adath Jeshurun, the first Orthodox Jewish congregation in Boston.) Boston architect Charles Bulfinch would be pleased to see Dudley's new row houses, postmodern and pink though they may be. It was activist Bulfinch, after all, who promulgated the idea of an urban village. And come to think of it, his Tontine Crescent (see page 105) contained a Catholic church, Boston's first.

On foot, in a car, or on a bus, pause at the hub of Blue Hill Avenue and Dudley Street, or Hampden and Dudley Streets, and you'll see a modern version of old Boston—a church, a common, a hill, row houses and settlers. DSNI calls it the Urban Village. The community calls it home.

GOVERNMENT, POLITICS, & LAW

CHRISTMAS BAN

⚛ 1 6 5 9 ☙

It wasn't the Grinch who stole Christmas. It was the Puritans.
Christmas was banned by the Massachusetts General Court in
1659, just a few decades after the Massachusetts Bay Charter set
up the basic structure of government. With a brief lapse demanded
by the mother country, Christmas was a crime in Massachusetts
for two hundred years. Though opposition dwindled even as the
law stayed on the books, the ban was not removed until 1856.

A law against Christmas may sound fanciful or farcical or
quaint. But in seventeenth-century Boston it was serious business.
The first legal ban on a popular holiday illustrates the Puritan
agenda and their power, and reveals much about colonial mores
and practices—including the brief relief Christmas provided for
acting wild and crazy—and the underlying religious, social, and
political conflict between Boston's Puritan founders and the community they governed.

The ban was not a case of cranky old men tamping down good
fun. It was morality legislation, a major victory in the colonial culture wars. Puritan opposition to Christmas was fierce and had been
going on for decades, starting in England. The holiday was everything these reformists—some would call them zealots—disdained
and condemned: a raucous pagan revel that encouraged sloth, revelry, and licentious behavior. It exemplified what they had been
trying to cleanse from the Anglican church. In England the holiday was celebrated with ceremony and abandon (ceremony for the

royals, abandon for the riffraff, which is to say, mostly everyone). It was briefly outlawed in England during Oliver Cromwell's mid-seventeenth-century Commonwealth, when Christmas was declared *not* a day of feasting, but instead, of penance. The Boston Puritans were determined to get rid of it for good.

Christmas was a compromised holiday from the start—or from its start in Christendom. It was an ancient pagan holiday, associated with the winter solstice, and only later grafted onto Christian practices. Though the church had accepted it—hoping that religiosity would eventually trump revelry—Christmas originally had no theological association with Jesus's birthday. As Increase Mather put it in *A Testimony Against Several Profane and Superstitious Customs Now Practiced by Some in New England,* in 1687:

> In the apostolical times, the Feast of the Nativity was not observed.... It can never be proved that Christ was born on December 25.... It was in compliance with the Pagan saturnalia that Christmas Holy-dayes were first invented. The manner of Christmas-keeping, as generally observed, is highly dishonourable to the Name of Christ.

"Christmas-keeping" was, however, acceptable in other colonies, including New Amsterdam, Virginia, and Pennsylvania (though not by the Quakers), and a raft of non-English-speaking settlements. But in Plymouth, Massachusetts, the practice was also prohibited, though not by a legislative act.

In 1621, just a year after the settlers had docked, Governor William Bradford put a stop to this annual exuberance, this merriment under the guise of faith-keeping: Several young men had asked to be excused from their work, saying it went against their consciences to labor on Christmas. The governor, not wishing to impede piety, excused them, but then found them playing stoolball, a kind of country cricket, and other boyish sports. He called a halt, later noting in his journal: "If they made ye keeping of it mater of devotion, let them kepe their houses, but ther should be no gameing or revelling in ye streets." Though Governor Bradford's prohibition was verbal and not carried into law, it emanated from the same theological distaste as that of the Boston Puritans. The Boston moralists did not merely advise or counsel, but decreed, and set up a system of punishment: a fine of five shillings.

The ban was politically useful. In addition to legislating their religious views into law—not unlike governmental power plays in our own day—the Puritans sought to distance themselves from the Anglican Church and the mother country. Boston merchants, neatly merging commerce and religion (again, some may find echoes in our own day), refused to close their businesses on Christmas, achieving a zealous, three-pronged insult: to the Anglican settlers in Boston, the Anglican Church in England, and the Crown.

Though a ban may seem over the top even for the Puritans, the practices associated with Christmas make Mardi Gras actings-out seem tame. Celebrants reveled in orgies of food, drink, and sexual merriment—everything from cross-dressing and lewd displays to noisy, semipublic fornication (sex between the unmarried)—and a series of customs associated with reversing social status roles. Mobs of boys swept through the streets demanding food and drink from the affluent, vandalizing houses if their demands were unmet. The targets of their baiting and battery were the same individuals whom many of the revelers worked for during the rest of the year. Most prosperous households prepared elaborate spreads of food for the ruffians, in part to prevent damage to their property, but also to even the score, as they had exploited, abused, and humbled their hirelings all year long.

Some of the Christmas vigilantes were virtual gangs, with clever names and their own traditions. The Boston Anticks were hoodlum-thespians (perhaps yet another Boston First), who broke into the homes of the wealthy to stage ribald, unwelcome performances. If the pumped-up Anticks were denied entry, or served poor food or diluted drink, they tore up the property of their "hosts."

The thick catalog of customs associated with Christmas is so complex and varied, and so revealing of the societies in which they were practiced, that an entire genre of social history and anthropology has evolved. *The Battle for Christmas*, by Stephen Nissenbaum, though nominally a cultural history, reads like a delicious feature story in *Vanity Fair*—though more erudite and without glossy photographs.

Nissenbaum suggests that the holiday served a socioeconomic function, that it was a pressure valve that could be released after the hard work and necessary sacrifices of the harvest. Farmers were not only exhausted and starved for fun, they were angry and belligerent, having suffered for months under the yoke of their masters. Reversing social status, receiving gifts, engaging in revelry served to maintain the status quo.

Ban or not, there was a spectrum of opinion about Christmas. Some were opposed to its bogus association with Christ's birthday. Some felt that a holiday to honor Christ was acceptable, but did not wish to sanction uncivil behavior. Eventually, a posh social set in New York "invented traditions" for the holiday by adopting and adapting many foreign customs, including Santa Claus. These inclusions led to a more socially acceptable celebration, but one disassociated from spirituality. In the rising tide of Santa Claus, Christmas trees, holly, and mistletoe, the Boston ban on Christmas was finally lifted by the Massachusetts legislature in 1856.

Today, broadsides, blogs, sermons, and commentary are levied against the commercialization of Christmas—perennial calls to return the besmirched holiday to its pure, ecclesiastical origins. In fact, the day has no such origins. Christmas has always been a compromise between sacred and profane, the perfect holiday for human beings. Instead of banning it, the Puritans might have left it alone, if they could have foreseen a world of MTV and Internet porn, where there would be no need to condense licentiousness into one brief midwinter free-for-all. Today, much of this wild and crazy energy can be dissipated—Marxists would say sublimated—into shopping, precisely what Boston merchants who refused to close their doors on Christmas Day would have wished for.

MASSACHUSETTS STATE CONSTITUTION

❦ 1 7 8 0 ❦

The precedential and eloquent Massachusetts State Constitution (1780), model for the U.S. Constitution (1787), was created in two buildings, a grand public edifice and a small, modest home. It was written—standing, at a tall wooden desk—mainly by one man, John Adams, who had elements of both simplicity and grandeur in his character. He wanted to be a farmer, started to train as a minister, and instead became a lawyer, patriot, diplomat, vice president under George Washington, and the second president of the United States. His son, John Quincy Adams—who grew up in the house where John Adams wrote the Massachusetts Constitution—became the sixth president of the United States.

Through much of the eighteenth century, that grand public edifice—Boston's oldest public building, today called the Old State House—was the seat of royal government. The building housed administrative offices, records, a court, and post office. It became, unintentionally but inevitably, an incubator for liberty and the growth of democracy, the setting for the birth and ratification of the Massachusetts Constitution.

Before, during, and after the Revolutionary War, debate and disorder reigned inside and outside this sedate-looking edifice. Outside, the Boston Massacre occurred in 1770. Inside, in Representatives Hall, the debates of the day raged, particularly during the 1770s. Over time, the interior even morphed to the shape of representative government, a physical manifestation of political change. At one end was the council chamber of British royal governors, at the other a small court chamber, and in the middle, Representatives Hall, where the Massachusetts Assembly met, and where stirred the discontent of the colonists chaffing under British rule. A distinctly American adaptation was made, albeit before the existence of the USA: As more and more people showed up for debates, it was decided that a proper gallery was needed to seat them. In 1766, a gallery was built to accommodate citizens at Massachusetts Assembly meetings. This was a first in the English-speaking world, a public gallery as part and parcel of the legislative assembly.

The gallery was built almost fifteen years before the Massachusetts Constitution was ratified, but the exchange of ideas and escalation of demands that whirled about in this balcony helped lead to the Revolutionary War, to the creation of the state of Massachusetts and the United States of America, and to the writing of the Massachusetts Constitution and the U.S. Constitution.

During the 1770s, as the relationship with England was coming apart—was being ripped apart by the patriots—a new form of government was called for. It was called for, loud and clear, in the Old State House.

During the course of the Revolutionary War, energy was focused on winning, not on debate and writing. Throughout the war, Massachusetts was in governmental limbo. An early (1778) fledgling version of a constitution was rejected. In 1779, a committee was appointed, a First in itself, to set about the writing of a constitution. A few constitutions existed, including those of Virginia and Pennsylvania, but they had been written by the legislature and imposed upon the people without a consensus. In Massachusetts Bay Colony, where the Puritan heritage of community involvement was powerful, a body was appointed to draft a constitution for the people to review and vote upon. A subcommittee, consisting of

John Adams, Samuel Adams, and James Bowdoin, was instructed to write the draft. They in turn assigned the task to John Adams, who quipped—according to David McCullough's *John Adams*—that he was a "sub-sub committee."

Adams, a prominent lawyer—judicious, erudite, steeped in Democratic ideals—who had contributed to the Declaration of Independence, went to work. He composed his thoughts and the constitution in the small wood-frame house in Braintree (today Quincy) that he shared with his wife, Abigail, and their family. He worked in the same room in which he conducted his law practice. It was brightly lit—there were two windows, each with multiple panes of glass—and there were teenagers and young children about as he worked, as well as Abigail, who had earlier (in her famous letter of 1776) reminded him to "remember the ladies." In September and October 1780, he toiled at his desk—with paper and pen, passionate and practical ideas, surrounded by books. The constitution derived from the Magna Carta, the political thinking of the ancient Greeks, and the work of thinkers and theorists such as Hobbes, Rousseau, and Locke. John Adams was helped here and there by his radical cousin, Samuel Adams, and James Bowdoin. But Adams mainly wrote alone, putting the document together in four weeks' time.

What he wrote was an expression of the democratic ideals the patriots had debated and were fighting for, translated into a system of government. He drew on the state constitutions of Virginia and Pennsylvania, on the great works of the Enlightenment, and on his own *Thoughts on Government*, written five years earlier at the onset of the Revolutionary War. The language in his *Thoughts* is stately, cadenced, and restrained: "Kings we have never had among us. Nobles we never had. Nothing hereditary ever existed in this country; nor will the country require or admit of any such thing."

Above all, John Adams drew on the legacy of the Puritan settlers, the belief that preserving liberty depended, as David McCullough summarizes, "on the spread of wisdom, knowledge, and virtue among all people." Knowledge and learning are mentioned repeatedly in Adams's document; Christian leadership is assumed. The idea of the social contract is a dominant motif. "The body-politic is formed by a voluntary association of individuals," he says in the Preamble. "It is a social compact, by which the whole people covenants with each citizen, and each citizen with the whole people, that all shall be governed by certain laws for the common good."

The intent and shape of the document are familiar to anyone who has the slightest knowledge of the U.S. Constitution. There is a Preamble, followed by a Declaration of Rights. The arrangement of government calls for executive, judiciary, and legislative branches. The legislative branch is two houses, a Senate and House of Representatives. A governor is to be elected at large.

The document contains the ideals and purposes of representative democracy, and some of the personal philosophy of John Adams, who was surrounded by books, fed and led by their ideas. When visitors to Adams's home today see the modest but cheerful room in which he wrote, the shafts of sunlight in the air—creating a space for contemplation, a pause to let ideas float, merge, and bind, or dissipate and fall away—it is possible to imagine him composing, stopping to study a text, consider its meaning and application, as sunlight moves across the walls. One can imagine the weeks of thinking and writing as September turns to October, Indian-summer light flickering thin and gold. What is almost impossible to imagine is that he wrote the constitution in four weeks' time.

As representation was the essence of the new nation, Adams and the other patriots-turned-legislators felt it was essential that the draft be well understood and agreed to by all enfranchised citizens. Behold, then: in total absence of mass media and the Internet, and without wishes for either, the document was disseminated and debated in every hamlet, village, and town of the Commonwealth of Massachusetts. Heated debates took place in newspapers, in public gatherings, in meeting halls, and in churches, as ministers addressed the issues, framed the debate, then let congregants cut loose.

Never before had a blueprint for government been organized in this fashion—with free men reading the draft of the document, debating it in parlors and pews, and in the press, suggesting further sections, revisions, redos, and then voting upon it. "Townspeople did not blindly defer to their leaders and shower their handiwork with accolades," notes Richard Brown in *Massachusetts: A History.* "In town after town, a querulous individualism marked the evaluation of the document."

Replies were by county. The roll call and some sentiments are recognizable: Barnstable, Berkshire, Bristol, Dukes, Essex, Middlesex, Nantucket, Hampshire, Worcester, Suffolk (Boston, Brookline, Dorchester, Braintree, Stoughton), and Plymouth. Berkshire country registered its complaints (testily) about dominance by Boston.

In October 1780, by a narrow margin, the constitution was adopted. Within weeks, John Hancock was elected governor. The royal lion was gone from the great edifice that once housed the seat of royal government, and the building renamed the State House. King Street was redubbed State Street. Governor Hancock moved in.

It was a small world, a village of legislators, ministers, farmers; Hancock and Adams had been schoolboys together in Braintree, both the sons of clergymen. John Hancock's father had baptized John Adams in his boyhood. The boy, now the man, author of the Massachusetts Constitution, lingered briefly in Boston and Braintree following the ratification of the document and then returned to Paris to complete the peace negotiations between the new nation and England, leaving Abigail and his family once again. He had written the constitution, the Puritan-inspired compact, before the war was formally over, though in some ways it had been written years before, in 1620, aboard the *Mayflower.*

This being Massachusetts, the architectural "stages" for the first constitution still stand. The Old State House on State Street is a fine museum, run by the Bostonian Society. John Adams's modest house is part of Adams National Historical Park in Quincy, which also includes the mansion called Old House that John and Abigail later occupied, and there is an adjacent freestanding stone library with fourteen thousand books in thirteen languages. A short drive from Old House and the library are the salt-box house where Adams was born and the adjacent house that he and Abigail lived in, starting as newlyweds—once he had made improvements to the kitchen—and in which he wrote the constitution, standing at his wooden desk.

He did good work. We are still using the Massachusetts State Constitution. It is one of the first ever written, the first to have a convention called to write it, the first to have a populace ratify it, and it has survived basically intact for over 225 years. It is the oldest living, working, written constitution in the world.

GERRYMANDER

Gerrymander, gerrymander, gerrymander. Newspapers, radio, and television continually bombard us with this word, which was cleverly and roguishly coined in Boston two centuries ago for a rogue deed done here. Earlier, the deed had been committed elsewhere, but halfheartedly, not with the flair and flagrancy of Boston legislators. Editorialists and wags pounced on this opportunism-con-brio with a zeal particular to Boston. It takes a certain intellectual climate to appreciate and mark deceit, to deliciously confirm one's low opinion of others, to publicize and protest. A bad deed well named is more outrageous, reprehensible, and fun.

Like the pompadour and the petri dish, items named for their inventors—of a hairdo and a container for microbes—the gerrymander is named for Elbridge Gerry (1744–1814), a signer of the Declaration of Independence, governor of Massachusetts, and vice president under James Madison.

To gerrymander (verb) is to rearrange political districts to win votes rather than to accurately represent voters. A gerrymander (noun) is an official elected as a result of this maneuvering. By arranging the map of a district, batching or isolating citizens and their votes, electoral results can be skewed. Gerrymandering is an undemocratic practice that requires a democracy in which to work.

A brief primer on the origins of political parties in the U.S: During George Washington's second term, political parties began. The Federalists, led by Alexander Hamilton and John Adams, believed in a strong central government and power concentrated in the hands of the wealthy and well educated. The Republicans (later called Democrats) followed the philosophy of Thomas Jefferson, who believed that the government that governed best governed least, and that power should rest among all the people.

In 1812, Elbridge Gerry was in his second term as the Republican governor of Massachusetts. The Massachusetts state legislature, largely divided between members of the Federalist and Republican parties, passed a controversial law. Seeking to win more seats, Republican legislators reconfigured a North Shore district to isolate Federalist supporters and concentrate and consolidate Republican voters. The earlier district had followed geographically and demographically logical contours. The new district was more intricate and connived; its outline made an exotic, fanciful shape.

The term "gerrymander" was coined in a newspaper office, where, as in our own time, chance encounters between wiseguys and wordsmiths can lead to ridicule and highjinks that yield useful criticism in the form of satire.

Benjamin Russell, an ardent Federalist, editor of the journal *Columbian Centinel* and later the *Boston Gazette*, had hung a map of the redrawn North Shore district over his desk at the *Gazette*, the better to study and critique it, and to work himself into a lather. As he studied the outrageously redrawn district, who should arrive, sauntering in, but Russell's friend, portrait painter Charles Gilbert Stuart (1755–1828), well known for his rendering of President George Washington. Stuart, by then a legendary portraitist, was not above the art of the cartoon. He, too, was infuriated by the opportunistically redrawn district—an elaborate shape that snaked west from Salisbury and Amesbury, south to Andover all the way down to Chelsea, and then abruptly and improbably east to Marblehead. The shape, so obviously contrived, begged to be mocked. To Stuart, it looked irresistibly like a dragon. Encouraged by his pal the editor, Stuart sketched it, adding clawed feet to Salem and

Marblehead, a tail to Chelsea, and a heraldic wing to Methuen; Salisbury became a reptilian head with fangs and a savage tongue. Stuart drew quickly, deftly, gleefully, and stood back to admire his handiwork, relishing the mischief it would cause.

"That will do for a salamander," he quipped. Editor Benjamin Russell took one look and said it should be called a "Gerrymander." The men laughed. A fine satiric editorial, probably by Benjamin, was produced and published in the *Boston Gazette*, March 26, 1812.

The headline read: "The Gerry-Mander," and then, "A new species of *Monster,* which appeared in *Essex South District* in Jan. 1812." The cartoon showed a vengeful monster with arrowhead tongue, clawed feet, and bellicose wing. Below the drawing, the caption (from the Gospel of Matthew) read: "O generation of VIPERS! who hath warned you of the wrath to come?" The lengthy satire began, "The horrid Monster of which this drawing is a correct representation, appeared in the County of Essex, during the last session of the Legislature. Various and manifold have been the speculations and conjectures, among learned naturalists respecting the *genus* and origin of this astonishing production." The assessment goes on to examine the origins of the monster: "He [a Doctor Watergruel] therefore ascribes the real birth and material existence of this monster, in all its horrors, to the alarm which his Excellency the governor and his friends experienced last season, while they were under the influence of the Dog-star & the Comet...."

The practice of what we now call gerrymandering did not originate in Boston. Other examples exist in early American history. In 1709, representation of Bucks County and Chester County in Pennsylvania was rigged to prevent Philadelphia from getting its proportional piece of the pie. In 1732, George Burrington, royal governor of North Carolina, divided the precincts of his province to secure his power. Taking the dim view of human nature, which rarely disappoints, it is likely that since the dawn of representative government, someone has tried to thwart the process. When even

a pie is served to a family of eight—an even number, readily carved
—one sibling, perhaps destined to earn his living in a particular
way, will wheedle a larger wedge, and often with great charm.

But the cultural climate of colonial Massachusetts was fertile
ground for political shenanigans. Democracy was relatively new as
a system of governance, and without a lot of safeguards. The English tradition of dry wit, caricature, and ridicule was part and parcel of colonial literature. The press was coming into its day (see
page 43), and the great American tradition of political cartooning
was on the move.

In the good old, bad old days—mainly during the eighteenth
and nineteenth centuries—gerrymandering was generally employed
to amalgamate voters of a particular party. In recent years, as the
U.S. has become more diverse and electoral politics more complex,
gerrymandering has exceeded the fondest wishes of its early prac-
titioners. A district may be reconfigured to give advantage to ethnic, racial, or even religious groups, to rural vs., urban groups, or
to predominantly high-, low-, or middle-income blocs. In 2001
in Massachusetts, a great hullabaloo erupted as House Speaker
Thomas Finneran tried to gerrymander out of existence an entire
political district.

As data on voting habits and preferences becomes more volumi-
nous and precise—identifying voter preferences neighborhood by
neighborhood, block by block, even house by house—gerrymandering thrives. Still, many efforts of "singular ferocity"—to use the
term of Benjamin Russell, co-coiner of the word gerrymander—do
not succeed. The assault upon representative democracy by Speaker
Finneran was beaten back; he was unable to vaporize a district.
And two centuries ago, even following the ferocious maneuverings
of Massachusetts Republicans, Governor Elbridge Gerry was voted
out of office, one might say because of a word.

SCHOOL INTEGRATION

Roberts v. the City of Boston, the Legislative
Integration of Boston Schools, and *Brown v.
Board of Education of Topeka*

❦ 1 8 4 9, 1 8 5 5, 1 9 5 4 ❦

Those of us who learn just a little bit of history tend to regard events in isolation. A civilization falls, *kerplunk*. A war begins because of a few incidents. An age of enlightenment dawns—ta-da! —with the unscrolling of a document. As we dig into a subject, the long, slow, complex evolution of events becomes apparent.

But the story of *Brown v. Board of Education*, the 1954 Supreme Court case that ended legal segregation of schools in the U.S., is such a dramatic event that even those knowledgeable in American history may be forgiven for viewing it as a twentieth-century affair. The decades of legal segregation in the American South, discrimination in housing and employment in the North, and the continuation of segregation in the military all through World War II —known to even the most history-averse among us—appear to be more than enough background for *Brown v. Board of Ed*. The biographical drama of Thurgood Marshall—the NAACP civil rights lawyer, lead counsel for the plaintiffs, who would become the first African American Supreme Court justice—is a story unto itself. It is almost as though there is no room for additional expository material in *Brown v. Board of Ed*.

But this landmark case is in fact an incarnation, a continuation and manifestation of an earlier case. A legal battle in Boston a century before the historic Brown decision set in place almost every element of the events leading up to the 1954 decision. In 1849, *Roberts v. the City of Boston*, a matter of school segregation, was on the docket. In 1850, the Massachusetts Supreme Judicial Court ruled against the integration of Boston schools, indelibly inking into law the concept of "separate but equal" schooling, the idea that black and white schoolchildren could be educated separately with no harm to either. This decision, written by Justice Lemuel Shaw during the early years of the Abolitionist movement, created the legal justification for segregation in Boston schools. It also energized and formalized the movement toward equality in the African American community, and helped to forge an alliance between liberal whites and activist blacks that became a national model that forced social, political, and judicial change.

Think of Boston as a foundry. Events on Beacon Hill in the early nineteenth century led first to the denial of equal rights to black and white schoolchildren by a Massachusetts court; later to equal rights in Boston, by an act of the Massachusetts legislature (while school segregation was still practiced in other major American cities); and much later to the passage of a national ruling by the U.S. Supreme Court that outlawed segregation in American schools.

Strictly speaking, the case began with Benjamin Roberts and his daughter, Sarah, in Beacon Hill's black community during the early 1800s. Roberts wanted his daughter, age five, to go to school near home. It wasn't as though there was no school for Sarah. The Abiel Smith school, the "black school," had existed for several years, but it wasn't as good as the white schools, which had better supplies, better-trained teachers, and secure, attractive buildings. Black children living on Beacon Hill saw these shipshape white schools every day; they walked past them on their way to their own scruffy school. Sarah walked past five white schools to reach the only school she was allowed to attend.

Less strictly speaking, the case began because of the grievances among Boston's black citizens, who in spite of their poverty and poor education, agitated, strategized, and organized. By 1850, about half the city's black population lived on the north slope of Beacon Hill. The red-brick African Meeting House at Smith Court on Joy Street, built mainly by black artisans back in 1806, was a virtual black Faneuil Hall, a combination church, meeting house, school, and community center, and also the recruitment center for members of the Massachusetts Fifty-fourth Regiment in the Civil War (see page 219).

The existence of this concentrated, close-knit black community fostered the move toward equality. During the early 1800s, public education was a major concern in Boston; the first public high school in the country, Boston Latin School, for white children only, had been established back in 1645. Black Bostonians, aware of the public school movement even as it was denied to them, knew that education was critical. In 1798, Primus Hall, who had fought in the Revolutionary War and was a leader in the black community, set up a makeshift school in his Beacon Hill home. In 1808, an actual school, though not a public school, was established in the basement of the two-year-old African Meeting House, and was greatly aided by an endowment from Abiel Smith, a white merchant, in 1815. (Both the school and meeting house still stand and are part of the Museum of Afro-American History.) An adjacent building, the Abiel Smith School, evolved from the basement school. When it was admitted into the Boston school system in 1835, its black teachers were fired.

Just as Arthur Miller wrote a play about the Salem witch trials that was also about McCarthyism in 1950s America, two authors, Stephen Kendrick and Paul Kendrick, wrote a book about the Roberts case that has powerful flash-forward references to *Brown v. Board of Education*. *Sarah's Long Walk: The Free Blacks of Boston and How Their Struggle for Equality Changed America* is a work of history, not historical fiction. But its uncovering of the wellsprings of school segregation is so assiduously researched and chronicled as to reveal the evolution of legal equality—Sarah's metaphorical long walk —as inexorable. The authors uncovered and relate the long lives of men and women focused on social change.

In 1847, Benjamin Roberts, a printer, educator, activist, and journalist—founder of the *Anti-Slavery Herald*, a newspaper of, by, and for the black community—tried to enroll his daughter, Sarah, in a white school close to home. The Otis School accepted her. Within a few months, the Boston School Committee got wind of this infraction, and sent a police officer to remove Sarah, age four, from her classroom.

This was not to be the end of the matter. Benjamin Roberts received his distraught daughter. She was already marked by history. Little Sarah was named for her grandmother, Sarah Easton Roberts, who as a child growing up near Bridgewater, Massachusetts, had sat with her family on the floor of a white church to protest segregation, after finding the pew they had bought first tarred, then pulled down.

Flashback to Thanksgiving Day, 1836: In a grand house on Essex Street in Salem, Massachusetts, a young servant, Robert Morris, met Ellis Gray Loring, scion of a Brahmin family. Loring, a lawyer in his twenties, was reform-minded, freethinking. Morris was a "table-boy" in the house of John King, a Salem lawyer and friend of Loring and his wife, Louisa. The teenaged Morris and twenty-something Loring met over bread at Thanksgiving dinner, though they would not break bread together for several years. As Morris served, Loring noted his alacrity, affability, and poise. He took him on as his own servant, grew to admire and respect him. When Morris was of suitable age, Loring asked, according to *Sarah's Long Walk*, "Do you wish to learn a trade, or do you wish to study law?"

Loring, by then a prominent abolitionist, one of the founders of the Anti-Slavery Society, became a mentor to Morris. Morris, who was both gifted and driven, became one of the first black lawyers in the United States, and the first to win a jury case. His services were sought by black and white alike.

Meanwhile, Charles Sumner, a respected attorney, was nosing about for something important to claim his heart and mind, as his energies and élan were insufficiently engaged by his legal practice. On the surface, he seemed yet another estimable white man, a fellow of privilege. But his origins were tougher, less sheltered, closer to those of the workingman. His father was a liberal sheriff, and Sumner had grown up side by side with black neighbors who lived on the same side of Beacon Hill as he and his family.

Benjamin Roberts seethed with indignation on the expulsion of his daughter, Sarah. He believed that she was entitled to attend any school, as a freedom granted by the Massachusetts State Constitution (see page 201), and sought legal counsel from Robert Morris. Morris agreed to help, but as he had passed the bar just seven months before, he invited Charles Sumner, a practiced and respected jurist, to appear with him in court. Unknown to Morris, Sumner was looking for a case worthy of his passion.

Sumner and Morris were a phenomenally effective combination, not only because of the novelty of a black and white legal team— probably the first interracial team to argue a brief—but because their skills and styles were complementary. Sumner—the learned, wise, and seasoned jurist, philosophical in temperament—fashioned an elegant statement that bound together rationales ranging from the principles of democratic government and Christianity, to current ideas gathered from the local black community, to the rousing writings of William Lloyd Garrison's abolitionist newspaper, *The Liberator*. Morris—young, fresh, street-smart, and passionate by temperament—powerfully summarized Sarah Robert's attempts to gain admission to schools close to her home and the damaging and unconstitutional rebuffs.

In 1850, in a unanimous opinion drafted by Chief Justice Lemuel Shaw, the Massachusetts Supreme Judicial Court ruled unanimously against the plaintiff. But Boston's black community had been mobilized and organized. School segregation continued as a public issue, with additional attention focused by rising tensions between the North and South, and multiple divisive incidents—including the ugly recapture of escaped slaves peacefully employed in Boston—that presaged the Civil War.

In the end, Boston schools were integrated five years later by an act of the Massachusetts legislature, rather than a judicial decree. Reformer Charles Slack, a state representative from Boston who became chairman of the Joint Committee on Education, prepared a report for his education committee, spurred by a grassroots petition drive. Not only was his report based on the Roberts case, but the petition grew directly from activism in the African American community. There was heated debate, but the law passed with just six legislators opposed. It was signed into law by Governor Henry Gardner on April 28, 1855. The following September, black schoolchildren entered Boston schools. Boston had become the first major city in the country to integrate. A committee of mothers from the black community—mothers had been the mainstay of the activist movement—traveled from school to school to ensure that children were kindly and equitably received. Boston's activism—petitions, meetings, pamphleteering, boycotts, and organizing in the black community—became a national model, not only in the nineteenth century, but into the Civil Rights era of the 1960s.

Though Boston was the first major city to integrate its schools, the wording and logic of Judge Shaw's (1850) response to the Roberts case stayed on the books. Though Shaw's "separate but equal" rationale for segregating Boston's schools had been canceled by an act of the Massachusetts legislature, his language was precise, neat, and powerful. It would be carved into constitutional law by *Plessy v. Ferguson*, the 1896 Supreme Court decision that allowed separate transportation for blacks and whites, which extended into "separate but equal" restaurants, drinking fountains, swimming pools, streetcars, hotels and motels, theaters and movie houses, and public schools. The concept of separate but equal—though repeatedly challenged as inherently unequal—would linger for generations, until *Brown v. Board of Education* in 1954. Another African American father found it unbearable that his child would have no choice beyond the black school. Oliver Brown of Topeka, Kansas, had "had it" with watching his daughter Linda, age seven, forced to attend an inferior school, and he turned to the NAACP. The NAACP based its constitutional case—on behalf of several plaintiffs—on the Fourteenth Amendment's guarantee of equal protection before the law. (Boston's Charles Sumner, who became a senator, was a major player in the writing of the Fourteenth Amendment, note the authors of *Sarah's Long Walk*.)

Chief Justice Earl Warren's momentous decision, delivered May 17, 1954, cited the Roberts case. Jurists separated by a century wrote of the pernicious effects of segregation. Sumner had argued, "The separate school is not an equivalent." Chief Justice Warren wrote, "Separate educational facilities are inherently unequal." Sumner described black children as deprived of "those healthful, animating influences which would come from participation in the studies of their white brethren. It adds to their discouragement. It widens their separation from the community." Warren wrote that separate facilities created "a feeling of inferiority as to their [black children's] status in the community that may affect their hearts and minds in a way unlikely to be undone."

Years after the historic case challenging school desegregation, Benjamin Roberts, father of Sarah, said of his work with attorney Robert Morris and the black community, "The cause of equal school privileges originated with us. Unaided and unbiased we commenced the struggle."

It is a long struggle, Sarah's walk, not entirely won. As history teaches, an age of enlightenment does not soon follow the unscrolling of a document. But documents such as legal briefs and broadsides, and decisions made in legislatures and courts, honor those who have contributed to a cause and inspire those who carry on. The hundred-year-long walk from Sarah Roberts in Boston to Linda Brown in Topeka made equality in U.S. schools the law.

MASSACHUSETTS FIFTY-FOURTH REGIMENT

᙭ 1 8 6 3 ᙭

Nothing, be it societal or personal, happens in isolation, as some grand event that springs unfounded from nature. No friendship forms, no war begins, no discovery is made except in context. But some events rise above their context, revealing time, place, and circumstance—as though casting a beam through shadow.

These "beamlike" events exist through Boston history (and any history), and continue to illuminate. Many of the buildings and streets, institutions and attitudes that existed over a century ago are still before our eyes, on our minds, in our hearts.

The formation and fighting of the Fifty-fourth Regiment, the first all-black volunteer regiment in the Civil War, is a case in point. This was a remarkable enterprise in its time; examined today, it is more so. Led by a young, white, patrician colonel, composed of hundreds of black men from near and far, lauded by progressives—though its soldiers were initially paid less than their white fellow soldiers—the regiment illuminates the racial politics of the day.

Boston had been a hotbed of abolitionist sentiment for decades, an outgrowth of Puritan conscience and social morality (including that of slave-owning Puritans), liberal thinking, and the transcendentalist movement, which stressed social harmony and reform. Slavery was outlawed in Massachusetts in 1783. Harriet Beecher Stowe's *Uncle Tom's Cabin* was published in 1852. Boston integrated its schools in 1855, the first major city in the nation to out-law segregated schools (see School Integration, page 211).

By the time of the Civil War, John Andrew, an abolitionist, was governor of Massachusetts, and black abolitionist Frederick Douglass had become a prominent leader and speaker for the movement, along with many white leaders, including William Lloyd Garrison and Wendell Phillips. Black men and women, a part of the population since 1638 when they were brought as slaves, had organized, and continued to. There were black churches and schools, meeting houses, and an Underground Railway to guide Southern slaves to freedom.

Two years into the Civil War, Governor Andrew petitioned Secretary of War Edwin Stanton to form a regiment of black enlisted men in Boston. Five months later, a regiment had formed—scores of local African American men and hundreds from beyond Boston, including Canada and the Caribbean, volunteered. The ranks of this precedential infantry grew to over one thousand men.

The leader of the regiment, Robert Gould Shaw, was a handsome, privileged, intelligent, and idealistic member of one of Boston's most prominent white families, a family of thinkers and abolitionists. Just twenty-five years old and newly married, Shaw was Harvard educated and—unlike many wealthy young men of the day—had not hired a mercenary to take his place, but had himself enlisted in the Civil War and fought for two years in the battles of Virginia and Maryland. He advocated the use of black troops, which he referred to in a letter to his parents as an "instrument that would finish the war sooner than anything else." Massachusetts Governor Andrew asked Shaw to lead the Fifty-fourth. (The story of the regiment—its inception, evolution, battles, and effects—is worthy of a movie. *Glory,* starring Denzel Washington and Morgan Freeman, Matthew Broderick and Cary Elwes, directed by Edward Zwick, was released in 1989.)

Though the troops drilled mightily and fought valiantly, they did not fare well in South Carolina, where they had been sent to attack Fort Wagner at the entrance to Charlestown Harbor. A terrifying battle ensued, with the Union soldiers overpowered by the Confederates two to one. Shaw was the first to launch himself into the terror, climbing the ramparts of the fort. He was killed, along with many of his men. In an ugly, unsoldierly act, rebel troops completely stripped his body and tossed him into a pit with his men, the black soldiers he was devoted to, and they to him and to each other. While it was not intentional, dignity and reverence ultimately resided in this act. It was expected that Colonel Shaw's parents would exhume his body and bring him home to Massachusetts for burial as an officer and a gentleman. They refused, leaving his body where it was, the men of the Fifty-fourth Regiment buried together, resting in peace.

So much of the legacy of the regiment remains in Boston, included in the tangled racial history of black and white. The neighborhoods that surrounded the fabled corps still exist—Beacon Hill, home to Colonel Shaw's family and, at the foot of the hill, Boston's black community. And Readville, Hyde Park, where the regiment trained at Camp Meigs.

The 1806 African Meeting House on Beacon Hill, where Frederick Douglass helped to recruit the men of the Fifty-fourth, still stands. Built by free black laborers, it began as a black church and became a black Faneuil Hall, where abolitionists such as Frederick Douglass and William Lloyd Garrison spoke. Today the hub of the Black Heritage Trail, the venerable house is one of fourteen sites—not historic references, but actual buildings.

Beacon Street, where the regiment paraded before leaving for battle in South Carolina, looks eerily as it did in May 1863, when the proud regiment was cheered by a crowd of twenty thousand lining the street. The gold dome of the State House (1798) still gleams.

The regiment also seems alive because of the magnificent

bronze monument directly across from the gold dome, before which the newly formed regiment strode. Created by Augustus Saint-Gaudens, working with architect Charles F. McKim, who became his colleague on the building of the Boston Public Library (see page 117), the eloquent bronze is often described as the city's finest work of public art. Classical in design, depicting heroes marching to battle, brandishing rifles and swords, the bas-relief literally and figuratively reaches out to us. The work shows Shaw on horseback leading twenty-three of his infantrymen. The soldiers are archetypes rather than actual soldiers. In reality, two of the men were sons of Frederick Douglass, who actively recruited for the regiment. Sargent William Carney of New Bedford was the first African American winner of the Medal of Honor. Carney saved the regiment's flag as he narrowly escaped death during the carnage on the Fort Wagner battlefield. That torn but extant flag, representing the principles men died for, belongs to the Commonwealth of Massachusetts and is today stored in the state archives at Columbia Point, site of the JFK Library.

Saint-Gaudens's statue of the Fifty-fourth Regiment, shaded by the same two American elms—now towering, and with massive trunks—that it was placed between a century ago, is the first site on Boston's Black Heritage Trail.

In May 1997, a celebration took place in Boston: the centennial of the dedication of the memorial to the Fifty-fourth Regiment. African American Civil War enactors camped out on the Esplanade, a reenactment of the Civil War Veterans March of 1897 took place (reminding all that it had taken over thirty years after the Civil War to get the statue built), and the movie *Glory* was screened outdoors on the Charles River at the Hatch Shell.

In *Boston Sites and Insights*, her engaging book about Boston landmarks, author Susan Wilson includes Colin Powell's tribute to the men of the Fifty-fourth. Powell attended the 1997 centennial event in Boston. He later wrote about the deep significance of the

Regiment in his life, and its influence on his career, in a 2001 collection called *Hope and Glory: Essays on the Legacy of the 54th Massachusetts Regiment:* "To my dying day, I will never forget that I became chairman [of the Joint Chiefs of Staff] because there were men of the Fifty-fourth, Buffalo Soldiers, Tuskegee Airmen and others who were willing to serve and shed blood for this country."

SAME-SEX MARRIAGE

❦ 2 0 0 3 ❧

Thirty-two years is a long engagement, even for the most patient and loving partners. But Linda Davies, sixty-seven, and Gloria Bailey, sixty-two, continued their engagement not only because they loved each other, but because they couldn't get married. Same-sex marriage was illegal in Massachusetts and everywhere else in the United States—everywhere else in the world, in fact, except Belgium, the Netherlands, and the Canadian Provinces of Quebec, Ontario, and British Columbia.

In Massachusetts, even longtime same-sex partners—virtual spouses—have been denied the legal rights afforded by civil marriage. Infringements on ordinary liberties ranged from not being able to assert the rights of parents, even when they were legal adoptive parents or live-in partners of biological parents; to not being able to visit a critically ill partner in the hospital; to not being eligible for employee benefits for families, including healthcare. Other rights limitations involved property ownership, insurance coverage, tax deductions, and child custody. Less easy to codify and evaluate were the pleasures and pride of deepening a commitment and being married in the eyes of the community.

"We want to be married for the same reasons any people want to be married," was the refrain of gay couples, heard again and again —in sound bites on radio and TV, quoted in newspapers, and said quietly, by way of explanation, to family, friends, and coworkers.

In April 2001, Gay & Lesbian Advocates & Defenders (GLAD)

filed a suit on behalf of seven gay and lesbian plaintiffs—from cities and towns all over Massachusetts—that sought to remove the barriers to same-sex marriage, which GLAD charged were discriminatory. Each of the couples had applied for a marriage license and been denied. The defendant was the Massachusetts Department of Public Health (DPH); in Massachusetts, DPH enforces the state laws pertaining to marriage.

For two years the citizens of Massachusetts—and anyone else who chose to read, listen, and consider—were engaged in a great public debate on affairs ranging from the definitions of marriage and family, to the rights of courts to "define morality," to why gay lovers would *want* to get married, since so many straight couples had rejected the institution, especially in Massachusetts.

Three years into the new century, the Massachusetts Supreme Judicial Court (SJC) was the scene of a matrimonial Boston First.

On November 18, 2003, the court ruled that same-sex couples have the right to marry, to have the same legal rights, protections, and privileges of all spouses. The historic decision at the SJC—located in the heart of civic Boston, the executive, judicial, and legislative seat of government—meant that Massachusetts became the first state in the nation where same-sex couples were given the legal right to marry. The decision defined marriage as a fundamental civil right that cannot be denied to any resident of Massachusetts based on sexual orientation or gender. The court also ruled that the state legislature would have 180 days to change state law to comply with the decision.

The decision read, in part:

Recognizing the right of an individual to marry a person of the same sex will not diminish the validity or dignity of opposite-sex marriage, any more than recognizing the right of an individual to marry a person of different race devalues the marriage of a person who marries someone of her own race.

If anything, extending civil marriage to same-sex couples reinforces the importance of marriage to individuals and communities. That same-sex couples are willing to embrace marriage's solemn obligations of exclusivity, mutual support and commitment to one another is a testament to the enduring place of marriage in our laws and in the human spirit.

Linda Davies, one of the plaintiffs in the suit, heard the news on her car radio, en route to Boston from her home on Cape Cod. She proposed to her longtime partner, Gloria Bailey, who sat beside her. Bailey accepted, formalizing their engagement of thirty-two years. On May 17, 2004, the first day it was possible, Davies and Bailey applied for and received marriage licenses, along with hundreds of other couples all over Massachusetts, not only in Boston and the nearby cities of Cambridge and Somerville, but in sleepier settings, too, including scores of small town halls where tulips and daffodils bloomed outside in tidily mulched rows. For many days, newspapers, TV, and radio carried highly charged and moving reports of church and synagogue weddings and civil ceremonies in nondescript offices too small for family and friends. Marriages were celebrated in every imaginable setting, ranging from windswept dunes on Cape Cod to elegant hotel suites in Boston, to a Christmas-tree farm in Rowley. Impromptu parties were staged outside Boston City Hall.

"Your marriage is an example to others of how life is supposed to work," said Rosaria E. Salerno, Boston city clerk, to Joe Rogers, fifty-five, and Tom Weikle, fifty-three, a couple of twenty-five years. Reporter Pam Belluck described the event (*New York Times*, May 18, 2004). The longtime couple had camped outside Boston City Hall to be the first same-sex fiancés to be married in Massachusetts. "You really are already married," Salerno continued. "The only thing that's been wrong with your marriage, if I can put it that way, is that it hasn't been public."

During the years preceding the decision—as in past debates

over abolition, suffrage, the civil rights battles of the 1960s, women's rights, abortion, Vietnam—ordinary citizens became engaged in discussion. It was impossible not to. For two years, the issue was debated in the Massachusetts legislature (long before the issue formally belonged to them); on radio, TV, and in the newspaper; in homes and offices; in churches and synagogues; in every space, in every place possible, from the gym to the produce aisle to the gas station. Unlike many other political issues, this one made everyone feel free to weigh in. Who lacked a working definition of marriage, commitment, and love? And everyone had gay friends, coworkers, relatives.

To add additional nuance, Massachusetts's Republican governor, Mitt Romney, repeatedly and strongly spoke out against gay marriage, and legions of the state's Democratic legislators—many of whom are Catholic, and whose church is opposed to gay marriage —crossed ranks and sided with the governor. President George W. Bush did not refrain from commenting on this particular state issue. Massachusetts senator John Kerry weighed in as well, along with every other candidate in the 2004 presidential race.

Many citizens, including some who supported same-sex marriage, felt that the issue should be decided by the legislature, rather than through judicial interpretation of the state constitution. In the decision that favored the case for gay marriage, dissenting judges cited this concern in their reasoning. It is expected that a state constitutional amendment on same-sex marriage will be presented to Massachusetts voters in 2008. Its results will probably not affect those already married.

It was no accident that traditional marriage laws were challenged in Boston; the Commonwealth has a long tradition of extending civil liberties beyond decisions of the U.S. Supreme Court. The Massachusetts State Constitution (see page 201) included equal protection guarantees eighty-eight years before they were ratified as the Fourteenth Amendment of the U.S. Constitution. The state document grants broader rights than the federal on a variety of day-to-day matters. In Massachusetts, state-administered Medicaid pays for medically necessary abortions, and workplace drug testing is restricted well beyond federal legal limits.

The SJC's decisions—broader, more inclusive—continued during a decade in American history when the definition of family and "family values" had been used to exclude all but the most traditional families. In 1993, the SJC permitted gay couples to adopt children. In 1999, the court granted visitation privileges to gay partners who were "de facto" parents. In 2004, grandparents of children in families with gay parents were given visitation rights. "The court has redefined the legal definition of the family to comport with the practical changes we've seen, to make sure our legal definitions keep up with common practice," observed Paul Martinek, a lawyer and editor of *Lawyers Weekly USA*, in an article by Thanassis Cambanis (*Boston Globe*, November 19, 2003).

Like many Boston Firsts, the 2003 decision on same-sex marriage linked with other precedents; the Massachusetts Constitution was the first constitution in America, written in 1780, and a model for the U.S. Constitution. It guarantees that all people shall be treated equally and enjoy fundamental liberties. The Massachusetts Supreme Judicial Court ruled that to exclude same-sex couples from civil marriage was unconstitutional under the state constitution's equality and liberty provisions.

Chief Justice Margaret H. Marshall is herself a First—the first woman chief justice of the Massachusetts Supreme Judicial Court in its history, and only the second woman to be appointed to the court since its founding in 1692.

The day in May 2004 that gay couples were legally permitted to marry matched up with a great national first, as well: the fiftieth anniversary of the Supreme Court decision *Brown v. Board of Education.* This civil rights victory traces its ideological lineage to the Roberts case (see page 211) in Boston over 150 years ago: Sarah Roberts, an African American child, was refused admission at Boston's white schools. In 1849 her father, Benjamin Roberts, sued the city of Boston on behalf of his daughter, resulting in a Massachusetts Supreme Judicial Court decision against Sarah and the creation of the "separate but equal" idea, which reverberated through American life into the twentieth century, and which was finally overturned by Thurgood Marshall in 1954. This case was heard in the same court as the same-sex marriage case.

The recent decision leads to the persistent, unanswerable, but nevertheless fascinating question of why so many Firsts, especially those associated with liberal ideas, occur in Boston. Each new idea arrives and enters our open port as intriguing cargo. Our intellectual appetite, our market of ideas, cannot resist these new products. But each new idea is also regarded suspiciously and sometimes with hostility and derision, as though what has come before and is in place is immovable.

In our small but influential Northern city, we have long had our own war between the states—the state of being open, modern, and embracing, offering consideration, and the state of being closed, fusty, and intolerant, refusing consideration. The threat we see in change is that its newness will, like the advent of Boston Light, change the landscape. But the landscape does not change, only its illumination, its depths revealed by our ability to see.

NATURE & THE ENVIRONMENT

BOSTON COMMON

❦ 1 6 3 4 ❧

Watch your step as you try this exercise, which is best done with eyes semiclosed. (It's worth a little danger and a few odd looks to time-travel.) From the foot of Beacon Street, step down into the vale of old meadowland that is Boston Common. While following the curved bucolic path from Beacon Hill to today's downtown, let the present-day bustle slip away, and enter the protected greensward. Mentally, feather-in small reddish cows, as depicted in the landscape paintings of early New England, and some English variety of sheep, and diminutive, continental pigs—furrier, less pudgy and pink than those of today—and perhaps a scattering of goats. Let these outlines solidify and the colors intensify like images arriving on the Internet. Erase the skateboarders, dog walkers, men and women in suits, and speedwalkers clutching glinting metal hand weights. You will have a sense of the Common as it was when created by early governors almost four centuries ago, public land meant to serve and bind together a community.

The origins of what is generally considered the first public park in America are ancient and gnarled, and might best be depicted in a series of woodcuts. Native Americans first cleared the land. In 1634, a weary but determined John Winthrop and his Puritan flock arrived on the windy expanse seeking a better place to live—especially better water so they could brew their essential daily drink, beer. They had already tried an encampment in Charlestown, where water had been a problem. Newly arrived on the expanse—cleared land with hills, mainly deserted by the Native Americans—they found a self-sufficient, solitary roamer, Reverend William Blackstone, in rough residence on the meadow. The Puritans took stock, not to mention possession, gave the Reverend Blackstone fifty acres for continued roaming, and settled their tribe. Soon after, bachelor Blackstone, disinclined to live near others, sold most of his fifty acres to the new arrivals. The Puritan leaders recognized the value of shared land; Blackstone's former property was set aside for purposes of pasturing animals, parading and drilling militia, and executing criminals and sinners (a practice that doubled as entertainment). Pigs had their consigned eating and living area, as did sheep and cattle. Over time, a poorhouse was erected, and a graveyard called South Burying Ground; later the Old Granary Burying Ground was established in 1660.

Another graveyard, Central Burying Ground on the Boylston Street side, was added in 1756 to relieve the overflow, which is not a figure of speech, as some corpses in the multistacked Granary were floating about underground. Artist Gilbert Stuart, who commemorated George Washington—and drew the celebrated cartoon of a "Gerrymandered" Massachusetts district (see page 207)—is buried at Central Burying Ground, along with hundreds of British and American soldiers killed in the Revolutionary War. Earlier graves of Native Americans—four hundred of them, according to author-historian Susan Wilson—were discovered during archeological digs during the late twentieth century.

In the beginning, Boston Common was neat, open, and orderly —a combination pasture, market, and outdoor meeting place. But its acres became messy, an offense to the eye and the nose, and to propriety, as settlers increased in number and used the land as a dump. The tidy Puritans passed a concise, communal law specifying that "no person shall throw forth or lay any entrails of beasts or fowls, or garbage, or carrion, or dead dogs or cats, or any other dead beast or stinking thing, in any way, on the Common, but are enjoined to bury all such things so they may prevent all annoyance unto any."

The soil of this first public green, or commonage, as it was known, was fertile for other Boston Firsts. So many occurred on these forty-eight acres as to conjure up a tale for a children's book that night be titled *A Garden of Firsts*.

In what is today known as the Old Granary Burying Ground—next to Park Street Church—lie some of Boston's earliest settlers, including the first colonial arrival, a small girl named Ann Pollard. She set her dainty foot upon Boston soil when it was still known as Shawmut, not yet renamed by the Puritans in 1630. For many years, Pollard ran the Horse Shoe Tavern on the east side of Tremont Street, apparently to good effect. She lived to be 105 years-old. While a mere 103, she sat for a portrait, now in the collection of the Massachusetts Historical Society, and open to view. Boston's first mayor, John Phillips, is also in the Old Granary Burying Ground, along with Benjamin Franklin's parents—Ben decamped for Philadelphia—and Paul Revere, Peter Faneuil, dozens of Revolutionary War soldiers, and the five men who fell in the Boston Massacre, including Crispus Attucks, an African American former slave, the first man to die in this opening round of war.

Visitors today, treading through the hallowed ground in bright sneakers and colorful Polarfleece windbreakers, are fascinated by the alternate views of reality depicted on the rows of seventeenth-century headstones, which to the Puritans represented not asymmetry, but continuity. The soft-gray and mottled-white headstones are carved with leering death's heads and sensuous fruits of paradise. These are Puritan signs and signatures, "symbols of the departing soul and the blissful eternity that is its destination," according to historian Robert Booth.

In 1737, a large granary was built on the edge of Boston Common near the graveyard that later took its name. It was used for many years to supply corn and wheat to the indigent, but became

NATURE & THE ENVIRONMENT

ramshackle and infested by mice and weevils, and was sold to a commercial enterprise. In this facility, a sail-loft was built. Here the first sails for the first ship of its type, the USS *Constitution* (see page 113), were assembled and sewn by hand.

When the granary had become decrepit, it was torn down and in 1809 replaced with one of Boston's most beloved church buildings, graceful brick Park Street Church, designed by Peter Banner. This was a true Puritan congregation, a conservative "Trinitarian evangelical religious society," in stark opposition to the Unitarian advance upon Boston. Here, beneath tower, steeple, and belfry, behold yet another bevy of Firsts: On a Sunday morning in the autumn of 1895, the pastor was preparing his evening sermon, only to have a fusillade of mud and an avalanche of stones come crushing through window and wall, abrading and almost drowning him, smashing his desk to splinters and shards, and destroying the sermon he labored on. Though the source of the chaos seemed demonic and the minister proclaimed it so, it was actually the by-product of a workman who had been digging a tunnel for—but, of course—the world's first subway system (see page 85), which chugs along beneath Boston Common. Working on the Sabbath just outside Park Street Church, the whistling workman had broken a water main. Rising from the chaos to deliver his sermon that evening, the minister fulminated against sin, sinner, and subway, condemning the great project as "an infernal hole."

In his guidebook, *Boston's Freedom Trail*, illustrated by sketches, Robert Booth describes a slew of additional nineteenth-century Park Street Church Firsts: On July 4th, 1829, William Lloyd Garrison, the abolitionist and publisher of the *Liberator*—so influential in the Roberts school segregation case (see page 211)—gave his first public antislavery oration there. Three years later, the elegiac hymn "America" was first sung—introduced to America—by the church's children's choir. During a strike by Boston police in 1919, opportunistic Bostonians gathered on the edge of the Common to hold the world's largest craps game. As the police could not be called up, the revel was loudly broken up by the state militia, who arrived at the revel outfitted with bayonets.

On the Beacon Street side of the Common, across from the State House, the grand statue of the Fifty-fourth Regiment (see page 219) stands. Here black troops marching to serve in the Civil War paraded with the American and Massachusetts flags. The ancestors of some of these black men had grazed cattle in the Common and fished at the water's edge near today's Charles Street. Crispus Attucks, the black Bostonian shot in the 1770 Boston Massacre, is among the dead in the Old Granary.

As Boston grew and urbanized, cows, the raison d'être of the Common, were nevertheless banned from their longtime grazing grounds. The pasture morphed to park; paths and more defined places for public gathering were put in place, and trees. Still, the aspect of a meadow lingered. Charles Bulfinch adored this karma-rich public space, with its literal and figurative layers of history, and by 1815 had built a series of brick row houses on three sides of the park. During the later nineteenth century, the Common was incorporated into Frederick Law Olmsted's chain of parks known as the Emerald Necklace. As befits America's first public park, listed in the National Register of Historic Places, this bead is the first in Olmsted's chain, connecting next to the Public Garden, an elaborate and idealized idea of nature as opposed to a place used by man and beast for four hundred years.

In one of Boston's many echoes, its refractions and reflections back and forth across time, the modern appearance and uses of Boston Common are versions of its first self. It is still public, still green, still a gathering place for political expressions and public celebrations—everything from antiwar demonstrations to the welcoming of a pope. Public entertainment is still offered—Shakespeare in the park instead of the Puritan gallows and lopping off of heads (unless they are part of the play)—and the abundant trees, especially the scores of remaining English and American elms, all remind us of our arboreal past. Many of the original ponds are

gone, dried up or filled in, but the frog pond continues its existence, offering succor and fun to children during Boston summers.
During winter it is used for ice skating, surrounded by the few
gentle hills still suitable for sledding. In autumn the pond becomes
a reflecting pool, as it has since its beginning, or since any human
being or animal paused to gaze. Boston Common is the first site on
the Freedom Trail, gathering Boston's residents and visitors, past
and present uses, past and present lives. The first subway station,
Park Street, is here, across from Park Street Church, next to the
Old Granary Burying Ground, and near Bulfinch's remaining row
houses. We play out our days among them, and gather, now and
then, at the reflecting pool.

THE PUBLIC GARDEN

❦ 1 8 5 6 ❦

At a glance—which could be done standing in the middle of Charles Street—Boston Common is the venerable and dignified park; the Public Garden its younger, prettier sister. Boston Common was forest, then meadow, then town green, then a park, a changing place that evolved over four hundred years. The Public Garden is much younger and wasn't even a place at first, at least not for human beings, trees, and nonaqueous plants, as it was underwater. The land of today's pretty-sister park was marshland until the middle to late nineteenth century, when the soggy flats were filled.

Both parks are considered Firsts. The Public Garden is not, strictly speaking, the first botanical garden in America. New York's no-longer-existing Elgin Botanical Garden, founded by a civic-minded physician, Dr. David Hosack, came before (1801). Dr. Hosack's accomplishment, including the collection of thousands of plants, became high-rise Manhattan real estate—eventually Rockefeller Center.

But if civic involvement, grass-roots efforts, popular uprisings (mainly against real estate development), petition drives, proclamations, processionals of children (irresistibly attired on the Fourth of July), and squatters' picnics be the mark of America's first up-from-the-people public botanical garden, Boston's Public Garden it is. The Garden is yet another example of Boston giving rise to an annoying, insistent, indefatigable citizens' brigade, in this case lovers of horticulture, from pros to amateurs to converts. Invincible green brigades have organized in every age of the Garden's century-and-a-half existence.

The rather American idea of a public botanical garden has certain cultural requirements, a mix that combines and composts democracy, urbanity, love of plants, horticultural and botanical knowledge, and municipal support. And there needs to be land.

In the beginning, in Boston, there was no land for this purpose. Bostonians contented themselves with driving into the countryside, or visiting greenhouses at colleges and estates, or growing a bright geranium on a windowsill. But the plant lovers—many just a generation removed from living on the land—dreamed of something close by, easily reached, part of everyday life: a garden of varied trees, dense shrubbery, trellised vines, beds of colorful flowering plants, and perhaps even roses and palms in pots in summer. It would be a public place, this garden, open, accessible, and free. One could walk, sit quietly on a bench and read, admire nature, and learn from the labels that would be placed on trees, as in zoos, identifying and honoring each species, and citing its place of origin.

At the edge of Boston Common on (today's) Charles Street side was a beach—Benjamin Franklin, as a boy, fished there—and farther out were tidal flats from which a wayfarer or a beachcomber could catch views of Cambridge and Brookline. As the flats became more noxious and obnoxious, due to the motley disposal habits of piggish citizens, more enlightened and less sloppy citizens clamored to have the land better used.

Many machinations ensued, including the city of Boston giving away several acres of marshland, at the end of the eighteenth century, hoping private holders would clean it up. Later, as citizens agitated for plants, the city had to buy back from rope manufacturers the very land it had given away.

During the early nineteenth century, decades of squabbling began, as the value of land to be reclaimed from the Back Bay become apparent—including land that would become the Public Garden. The number of trees felled to provide newsprint to cover the issue could have filled a forest, let alone an urban park. The business community and revenue-hungry public officials sought residential development. A citizens' lobby claimed the land was public, guaranteed by the City Charter of 1822. The City Council called a meeting of all citizens in 1824; a representative from each ward was asked to prepare a report. This document is described in our own time by Henry Lee in an essay in *The Public Garden*, an encomium by the Friends of the Public Garden. (The Friends, founded in 1970, are the current generation of activists; Lee has steered the group for many years.) The citizens' report denied the city's right to sell the land, demanding it be kept "open for circulation of air from the west for the sake of the health of the citizens." The land was snatched from the jaws of developers, but then languished for about ten years, as neither city nor state did anything much.

Enter Horace Gray, a wealthy manufacturer of iron, who was mad for plants. Gray raised quantities of grapes worthy of still-life paintings, and opulent beds of velvet-textured exotic flowers. (Not content with the sumptuous and extensive plantings at his homes in Boston and Brighton, he rented an old circus building near Beacon Street and turned it into a gallery of botanical specimens and an aviary of exotic birds, a natural history catalog of living plumage and plants, a true Victorian creation!) From 1837 to 1847, the kindly industrialist-horticulturist poured money into the garden site, reclaiming the flats and laying out a boardwalk edged by flowers and plants, including Boston's first poinsettia. Tragically, in 1847, Gray lost his fortune, and his precious conservatory and its contents burned, and the botanical garden project came to a halt. In 1853, the City Council reacquired the land, planning to turn the ruined dream into residential real estate.

In addition to the Public-Garden-in-process turning to mud, there was a legal mess—a snafu of competing rights among city, state, and mill owners who had built nearby dikes and roadways.

Citizens reorganized, sharpening their metaphoric scythes and shears. Horace Gray's influence had waned, but as though sent by a green goddess, along came Reverend Charles Francis Barnard, a South End pastor who had been a garden proprietor during Gray's era of influence. A biography of Reverend Barnard noted in *The Public Garden* claims that, above all others, "Boston owes its beautiful Public Garden to the foresight, the indomitable persistence, and personal audacity of Charles Francis Barnard." The publicity-savvy pastor organized annual Fourth of July children's processionals and flower sales in the Garden and generated much good press for its recreational and health benefits. He even constructed a greenhouse, and raised and sold flowers.

In this rising tide of popular support, the state legislature settled the land claims, allowing Boston to buy the necessary acreage. Behold, the Public Garden Act, passed in 1856, mandated that that land be forever dedicated to public use. In the late 1850s, a plan was purchased from George F. Meacham, a young architect, and the image of the twenty-four-acre Public Garden we know today began to take form: the pond (later called the lagoon), winding paths, granite basins with fountains, cast-iron fences and gates. Two able city employees, Boston city engineer James Slade and city forester John Galvin, shaped and grew the Garden from its early years, Galvin and his crew setting out trees, shrubs, flowers, and turf.

William Doogue, an Irish American horticulturist who became superintendent of Boston parks, was a great boon to the Garden, working energetically in the public interest for decades. Among many stylistic controversies played out during his tenure, Doogue was criticized for his bright floral displays and flamboyant use of tropical plants, the showy alien species beloved by park visitors. Charles Sprague Sargent, director of the botanically correct Arnold Arboretum, condemned these floral usages as aesthetically and horticulturally inappropriate. Doogue, superintendent of Boston's *Public* Garden, was undeterred. Henry Lee reports his re-

sponse to Sargent's critique in *The Public Garden:* "Harvard," said
Doogue, "like other institutions, has to stand sponsor for many a
blockhead who has passed under her portals."

The beat went on, with botanical and political cultivations and
upheavals in every age. Traditions grew like moss. In 1877 Robert
Paget, an English émigré, brought the Swan Boats to the lagoon.
A shipbuilder, Paget started his concession with common row-
boats, but then saw *Lohengrin,* the opera in which a knight crosses
a river in a boat drawn by a swan. Paget designed elegant, foot-
pedaled "swanboats." The beloved boats are still operated by the
Paget family. The Garden was botanically altered by Frederick
Law Olmsted and incorporated into the Emerald Necklace. Today,
it bears shade trees that testify to its long history—majestic
beeches, super-sized ginkgoes, graceful Japanese scholar trees, and
a small, precious number of American elms *(Ulmus americana),* per-
sisting in the face of fatal disease.

The Garden's trees represent not only landscapers' choices of
what will survive and look good, but references to Boston history
and civic associations. The many trees of Asian origin relate to the
city's long association with Japan, as does the sixteenth-century
Japanese lantern. (Interestingly, the lantern is made of iron, res-
onating with the contributions and occupations of benefactor Horace Gray, a manufacturer of iron.) The willow tree boughs that
drape over and into the lagoon look like scenes on blue-and-white
Chinese export porcelain. The continuing use of bright plantings
maintains the garden's Victorian heritage and the showmanship of
superintendent William Doogue, who served, as he put it, "public
taste and public pride." Today, as in the past, most of the Garden's
plant material is grown in sprawling city greenhouses in Franklin
Park, carefully and laboriously, much of it from seed.

As you wander through the Garden—its circuit of paths, the lagoon and bridge—consider not only its shrubs, trees, and flowers,
but fountains, statues, and monuments, which are like pages in the
scrapbook of Boston. They emanate from different time periods,

commemorate different heroes and events, and differ in style and provenance. Some were donated by individuals, and some came about through citizen-sponsored fund drives. The statue of Charles Sumner, the abolitionist attorney and U.S. senator who argued the nineteenth-century Boston school segregation case (see page 211) with black attorney Robert Morris, faces Boylston Street. The Ether Monument commemorating the first public use of ether anesthesia in surgery (see page 81) dominates the Arlington Street/Beacon Street corner. The bronze equestrian statue of George Washington at the Arlington Street entrance, a fine work by Charlestown native Thomas Ball (1819–1911)—visible from blocks away on the Commonwealth Avenue mall—reminds us that George Washington was not only an American hero, but a Boston hero. Just yards from this Arlington Street statue, surrounded by bright flowers in spring, General Washington sat on his horse on Boston Common, reviewing his victorious troops.

METASEQUOIA GLYPTOSTROBOIDES, DAWN REDWOOD

❦ 1 9 4 8 ❦

Talk about capital and its potential for appreciation—that venerable occupation and preoccupation of Boston bankers—imagine that something so tiny and obscure as a seed from central China should grow into something as massive, valuable, and globally coveted as *Metasequoia glyptostroboides,* dawn redwood! An initial investment of smarts, thrift, scientific know-how, and the collector's zeal for plants—as romantic and passionate as love, and as unpredictably fruitful—did it. *Metasequoia glyptostroboides,* the "living fossil" revived by Boston's Arnold Arboretum, is widely considered by plant scientists "the Tree of the Century."

A canny contribution of $250, vast scientific acumen, the academic network stretching from Boston to China at the turn of the twentieth century, and the expedition of a lone Chinese botanist resulted in the first rescue of a tree species thought to be extinct. The towering conifer with tens of thousands of feathery green needles now graces gardens around the world.

Metasequoia glyptostroboides, which calls to mind Tarzan movies and dinosaur epics, grows to astonishing dimensions—heights of over a hundred feet, a trunk diameter of seven feet or more—and is a relative of our California redwood. Its genetic lineage is so ancient that botanists considered the tree extinct, though reports of it occasionally trickled in, in the way of the ivory-billed woodpecker, an "extinct" bird that reappeared—flashing its pointed, pearly beak and red plume—and was scientifically documented in 2004. But sightings of the prehistoric redwood had never been confirmed. The tree was known only from fossil records. As with ancient and extinct animals, the remains of plants lay etched and encoded in rocks. No living versions were thought to exist.

In the bitter winter of 1941, a roving Chinese botanist discovered a curious tree in a secluded valley of central China. The bitterness of the winter was suddenly nothing. The sensation of cold dropped away. At first he thought he had found *shui-sung*, water pine. It took several years to collect herbarium specimens. Closer analysis at the Department of Forestry at China's National Central University suggested a botanical treasure, a herbaceous voyager from a distant world. *Metasequoia glyptostroboides*, a plant species that had grown 100,000,000 years ago, appeared to be alive—looking much as it had during an era when dinosaurs sheltered in its shade.

The graceful deciduous conifer—which is to say a tree that has needles and cones like a pine tree, but that sheds those needles in winter—had apparently survived in a few remote, isolated patches in China. It was revered by the local people, who considered it divine. They called it *shui-shan*, water fir. But they were unknowingly accelerating its extinction by using the tree's attractive wood for interior décor.

After World War II, in 1946, C.Y. Hsieh, a young Chinese botanist, assistant to Wan-Chun Cheng, Professor of Forestry at National Central University, was dispatched into the wilderness to locate the species. Hsieh endured great danger and hardship to locate the stand of trees and to collect specimens, which he modestly but triumphantly brought back to Professor Cheng.

The Boston-Cambridge connection extended its boughs to China. Professor Cheng, excited by Hsieh's remarkable cache of seed-cones, sent herbarium specimens to his colleague, Professor H. H.

Hu, director of the Fan Memorial Institute of Biology in Beijing, asking his opinion. As luck would have it, Professor Hu (1894–1968) was the first Chinese botanist to receive a doctorate from Harvard University, an alumnus so to speak, of Boston's Arnold Arboretum. He published a paper announcing the discovery of the "fossil-tree."

Together, Professor Hu—the Boston connection—and Professor Cheng approached the Arnold Arboretum. Cheng sent the dried flowers and fruit (rounded cones) of the conifer to Dr. E. D. Merrill, director of Boston's Arnold Arboretum, telling him of the newly discovered tree of the fossil genus. Hu wrote to Merrill, requesting $250 to enable Professor Cheng's assistant, C. Y. Hsieh, to return to Sichuan to collect seeds. (Then as now, the Arnold Arboretum was located in Jamaica Plain on land owned by the city of Boston and leased to Harvard; the "living collection" was and is part of the university's research and teaching missions. The 265-acre grounds are open to the public.)

Dr. Merrill recognized a good deal and wrote to a colleague:

> In general, for what it would cost to send one man from here [Boston] and cover his salary and travel expense I could maintain a dozen expeditions in China, and from each one of the dozen would receive as our share a 50-50 split about as much material as [would have been gathered by] the one man sent from here!

Funded by the Arboretum's $250, Mr. Hsieh, later Professor Hsieh, set forth alone on a perilous trip—at one horrendous juncture walking seventy-two miles—and collected the precious seeds. Each cone held multiple seeds—DNA messengers across time. Years later, Hsieh wrote a modest but moving account of one of his solitary journeys, threading though narrow mountain passes on the border between Sichuan and Hubei provinces, known for its treacherous trails, infrequent visitors, and ruthless robbers and murderers who beset hapless travelers. His words are beautiful and spare.

Finally, at dusk on the third day, I reached my destination safely. I set out immediately to search for that colossal tree despite hunger, thirst, and fatigue, and without considering where I would take my lodging. It was February 19th and cold. The tree was located at the edge of the southern end of a small street. In the twilight nothing was discernable except the withered and yellowed appearance of the whole tree. My excitement cooled.

"Am I to bring back just some dried branches?" I asked myself.

The tree was gigantic; no one could have climbed it. As I had no specific tools, I could only throw stones at it. When the branches fell from the tree, I found, to my great surprise, that there were many yellow male cones and some female cones [containing seeds] on the leafless branches. I jumped with joy and excitement!

Hsieh gathered the seeds in 1947. They were planted in the Arboretum's greenhouses in January 1948 and germinated quickly, many by the end of the month—infant descendents of fossil ancestors. The Arboretum also shared the seeds it had received with scores of other institutions in Europe and America. The ancient tree rooted in far-flung gardens and "took off" in the marketplace. Widely planted, the revived fossil was prized not only for its antiquity and dramatic rescue from oblivion, but for its astonishing appearance. *Metasequoia glyptostroboides'* needles are numerous, feathery, and soft—tree lovers often stroke them—while its red, sinewy trunk and limbs are majestic and strong. In winter the needles drop, emphasizing the tree's architecture.

This fruitful and felicitous Chinese American venture was

greatly facilitated by a preexisting relation between Boston's Arnold Arboretum and Chinese plant scientists, several of whom had studied at the Arboretum, including Professor Hu. Dr. E. D. Merrill, Director of the Arboretum, had become an expert on Chinese plants during his earlier years as a plant scientist in the Philippines. Among his colleagues and friends were the very Chinese professors with whom he cooperated during the *Metasequoia* adventure.

Every dawn redwood at the Arnold Arboretum is an offspring of this original, late-1940s endeavor. Some can be found across from the Hunnewell Building near the main entrance, planted near the Japanese katsuras. You will not have to walk seventy-two miles through bandit-ridden territory to find the trees, some now over fifty years old, though it would be a graceful gesture to pause and consider Professor's Hsieh's efforts and reverence for plants, as you regard the feathery green progeny of his *Metasequoia* seeds. And to think of Professor Hu (1894–1968) studying at Harvard, and Professor Cheng (1904–1983), conferring with Hu over conifer needles and cones.

In 1996, a quite elderly Professor Hsieh, who had gone on to become an expert in bamboo, visited the Arnold Arboretum, accompanied by his daughter and granddaughter. The scientist who had gathered the seeds appears in a photograph with the *Metasequoia* he rescued, the Arboretum specimen grown tall, the man small, though towering in achievement.

Today *Metasequoia glyptostroboides* is planted in parks, botanical gardens, and large-scale landscapes all over the world. It was rescued from extinction by an intervention—both modest and spectacular—by plant scientists in China and Boston, where seeds of the "living fossil" were propagated. Every modern incarnation of the tree—and there are hundreds of thousands all over the world—derives from the Arnold Arboretum's search, seed, and rescue mission, a First in plant ecology.

BOSTON HARBOR CLEANUP

The Longest, Largest Effluent Outfall Tunnel in the World

❧ 1 9 9 0 – 2 0 0 0 ❧

By the 1980s, Boston Harbor, the colonial harbor of legend, the nation's sentimental favorite, was filthy. It didn't even look like water anymore, but like some noxious semifluid. It stank, too, having become a repository of raw sewage, including human waste. To those observing from the surface, its waters no longer seemed oceanic but inert, commercial, a waterway for cargo, but not for life. Majestic oil paintings of the once grand harbor hung in Boston's museums, depicting azure waves and whitecaps that conveyed the freshness of clean salt water and bracing sea spray. These historic images had no connection with the bilge at Boston's edge.

The degraded harbor even became a national campaign issue. In the 1988 presidential campaign, George H. W. Bush dissed Boston Harbor, calling it "the filthiest harbor in America," taunting his opponent, Massachusetts governor Michael Dukakis.

When talk began about the urgency of cleaning Boston's near waters—a mysterious gray murk that most citizens didn't even realize was the end point for sewage (household, industrial, and runoff)—the nonengineers among us imagined a gigantic motorized strainer, a super-sized construction sieve that would repeatedly dip down to remove centuries of gunk, leaving the harbor pristine.

But even a little reading during those halcyon days of the early environmental movement taught nonengineers that the problem was not what was there, but what came in every day, including raw sewage from the likes of us. What was needed was better treatment of the waste that flowed and overflowed from the forty-three towns in metropolitan Boston, and the removal of the treated water far, far, far away from the harbor, where it would be dispersed into the deeper sea. The treated effluent had to be released into parts of the ocean that were not rich feeding grounds for fish, and be "whisked" into these deep waters so that it quickly diluted and diffused.

What was dreamed up and built to move the treated water was only one element in the immensely complicated, fifteen-year-long Boston Harbor cleanup—an organizational effort comparable to founding a new nation, a technologically proficient and efficient one. But this particular "device" was, and is, so precedential and effective, so beautiful in its own way and so colossal, that accolades are due for an amazing sight Bostonians shall never ever see, though they use it constantly. Buried deep beneath Boston Harbor, the effluent outfall tunnel is the largest single-entry deep-rock tunnel mined with a single shaft in the world. It is 9.5 miles long, 24 feet in diameter, sunk 420 feet beneath the sea at its starting point at the Deer Island treatment plant. A gigantic tube of moving water, the tunnel extends into Massachusetts Bay, carrying Boston's treated wastewater into deep, distant, active waters, where the Gulf of Maine current is strong. Every day, about 360 million gallons of Boston's wastewater travels through the tunnel.

A similar approach to digging and building a tunnel had been used in Australia (consultants were hired to help in Boston), but that tunnel was just 2.5 miles long. Charles (Charlie) Button, majordomo of the tunnel project, officially Director of Construction, Boston Harbor Project—a man respected for his ability to solve complex problems in the most direct way possible, and to stay calm while doing so—says that Boston's harbor tunnel needed to be as long as it is "to get to where it needed to go." In Australia, where the ocean floor drops off sharply, a tunnel just two and a half miles long was enough to get into deep water, to disperse treated wastewater safely and responsibly. In Boston the ocean floor is shallow for many miles; the tunnel had to be almost four times the length of its Australian relative to move into sufficiently deep water to effectively dispose of treated wastewater. To protect marine animals, the tunnel also had to terminate in a place unattractive to marine life—a sandy, stony area, dull and devoid of interest to fish, as opposed to one plush and alluring with vegetation. Unlike a conventional tunnel that begins and ends above ground, this marvel had to be cut into the ocean floor from one end, a "single-bore" tunnel.

The tunnel begins at the Deer Island water treatment plant, another marvel, and ends in Massachusetts Bay. Its interior is the same size as the Callahan Tunnel, which has carried generations of Boston motorists and their cars from the North End near Haymarket Square to Logan Airport and Route 1A under Boston Harbor. The outfall tunnel carries effluent, not people and cars, and there is room for a lot of effluent, which is now treated wastewater, as opposed to sewage, thanks to a decade of municipal, state, federal, and private lawsuits; federal, state, and municipal legislation; plus contention, discussion, wrangling, negotiation, mediation, and compromise beyond calculation and measure. Charlie Button says the wastewater that enters the tunnel after treatment looks like ordinary water. There are no solids. It is not murky, not tea-colored, but clear. The harbor does not smell bad anymore.

The logistics of building an almost 10-mile-long tunnel hundreds of feet beneath the ocean seem impossible. A subterranean work force—including the men and women known as "sandhogs," the tunnel workers—labored for a decade, 1990 to 2000. A diesel-powered railroad traveled underground. Electricity powered the tunnel machines; a power line from the South Boston Edison plant provided a terrifying and essential 125,000 volts. That's 125,000 volts for people and machinery traveling under water. As "the hole" was drilled (the excavation dug out 1.6 million cubic yards of bedrock), precast concrete segments were put in place, creating the cylindrical tunnel. These segments were stacked and piled everywhere for months. Each unit—six fit together to make a cylindrical section of tunnel—is 12 feet, 8 inches long (like a bowed slab of concrete), 5 feet wide, and 9 inches thick. "We want it to last forever," says Button.

Men and machines rode about underground, underwater, in the tunnel for many years. In September 2000, it went online, filling up with treated wastewater from the new, state-of-the-art treatment facility at Deer Island (one of two new plants). No one can ever walk through the tunnel again. No one can see into it again. But the water is traveling through, and along the last mile, at intervals, fifty-five diffuser risers, or pipes, are capped with gravity-powered gizmos—diffuser caps with spigot-like ports—that "whisk" the wastewater into the ocean, further diluting it. Boston Harbor has come back to life in an almost miraculous fashion. Even Charlie Button, a practical man, a civil engineer, uses the word "magic" now and then to describe the process.

How did the harbor get so bad, filthy enough to need major overhaul and the expenditure of 3.8 billion dollars? Essentially: people, politics, and ignorance. The population increased dramatically, water and waste management are expensive, and it was long thought that the ocean could absorb and convert infinite quantities of waste.

In the beginning, the seventeenth-century colonists handled their wastes one by one (think camping in the woods). Later, privies lined with stone or wood were built, their contents moved elsewhere—including, undelightfully, the cellars of homes—or carted off at night to be dumped into the harbor. The first sewers (not intended for human waste)—aboveground gutters, really—were dug in the later 1600s, and decades later, constructed underground. Cities and towns acted mainly on their own without regional planning. From 1877 to 1885, an impressive drainage system was constructed to divert sewage from the Boston area—some eighteen cities and towns—to Moon Island, off Squantum in Quincy Bay, where it was held for release with the outgoing tides.

Still, by 1919, pollution was forcing closure of clam beds. Not until around 1924 was there any thought to treating sewage, according to Eric Jay Dolin, author of *Political Waters*, an environmental history of Boston Harbor. Remarkable though it may seem, not until 1952 was the first primary wastewater treatment plant built—on Nut Island, for harbor-bound flow from the South Shore. In 1968 another primary wastewater treatment plant was built at Deer Island, to handle the remainder of metropolitan Boston sewage. By the early 1970s, the volume of sewage was literally overflowing both plants; the supposed water authority, the Metropolitan District Commission (MDC) lacked the funds and authority to respond. Untreated and partially treated sewage was fouling Boston Harbor, disgracing it, literally and figuratively.

The history of the recent, modern Boston Harbor cleanup—including rigorous treatment of waste before it even enters the harbor—began in 1982, when a man jogging on a Quincy beach got human excrement all over his sneakers. Horrified and enraged, William B. Golden, a city solicitor for Quincy, immediately walked in his feces-laden footwear to Quincy City Hall and demanded recourse. A few months later, Golden, representing Quincy, filed a civil suit in state superior court against the MDC (and later against other state water management agencies), for its violations of state and federal permits in discharging sewage into Boston Harbor. The suit charged that the Massachusetts Clean Water Act was being violated by the disposal of untreated sewage from the long overburdened treatment plants at Nut and Deer Islands. This suit led to the state's creation of the Massachusetts Water Resources Authority (MWRA), which the (federal) Environmental Protection Agency sued in 1985, along with the MDC, citing continuing violations of the Clean Water Act. Books could be written about this saga of sewage and suing, and several excellent ones have been, including *Political Waters*, by Eric Jay Dolin, and *Mastering Boston Harbor*, by Charles M. Harr (Harr, a professor emeritus at Harvard Law School, was the court-appointed special master in the case that led to the creation of the MWRA).

In its own way, the 9.5-mile-long tunnel—or gargantuan sewage pipe—from Deer Island to Massachusetts Bay, a modern Boston First, is as beautiful as Boston Light, as ingenious as the telephone, and as clever as shipping ice to India. It took ten years to build and cost $320 million dollars. Near the end of the project, two men, both divers, were killed in the tunnel when their independent air supplies failed. Work stopped for a year as the MWRA sought and implemented a new plan for pumping air into the tunnel, even though it was virtually complete and all the equipment had been removed. Two other men were killed during the long course of the tunnel project: a tunnel engineer and a pile driver working on a barge in the ocean, putting on the very last diffuser cap.

At Deer Island, which now holds a well-landscaped park as well as a treatment facility, brass plaques dedicated to the men who lost their lives were erected inside an old steam-pumping station. The men and their work helped to heal the harbor—to get the one-of-a-kind tunnel beneath the ocean floor, to move the treated wastewater away from Boston Harbor, and to allow nature to do her healing work.

The engineers on the project, not to mention the biologists, have great respect for nature's participation. Charlie Button, who supervised throughout and is still the point man on how the whole thing works, stands back from the vast array of construction, engineering, gizmos, and gadgetry—tunnels (connecting plants and islands), sludge digesters, storage tanks, compactors, treatment plants, and a fertilizer-making facility—and remembers the "snowy years that washed the harbor out," and the aesthetically pleasing way that gravity, not electricity, powers the fifty-five diffuser pipes and four hundred diffuser ports that whisk the water into Massachusetts Bay.

"It's a marvel of nature to recoup as she does," he says, citing not only the quality of the water, but "all the critters that have returned." The incidence of deformed flounder is dramatically down, along with the bacteria count and presence of toxic metals. In the spring of 2005—as though sent by central casting—a playful young humpback whale sashayed through Boston Harbor. Button points to the treatment facilities and the outfall tunnel and to the thousands of people who worked on the project as agents in what he calls the "harbor's rejuvenation." He is admiring and respectful of the ocean, and protective of it. He and his team used a 9.5-mile-long tunnel and the "assimilative capacity" of the ocean to do a job that civilization—from the scant scores of first colonists to the teeming millions of us—has imposed.

"All though history, people wanted the waste to go 'away,'" says Button. "There is no 'away.'"

There is only treatment, and the tunnel, and the visit of the whale.

Acknowledgments

I wish to acknowledge the enlightened individuals I have met through research and writing about Boston. I wish I could have had lunch with them.

Charles Bulfinch comes immediately to mind. At a leisurely pace, one facilitating observation and conversation, we would have walked together to the Tontine Crescent and quietly admired it. And then—perhaps I would have taken his worsted arm—we would have lunched on oyster pie with flaky crust and goblets of chilled Chablis. I would have liked to have called on Isabella Stewart Gardner, and sat with her amid the flowers of Fenway Court, the madonna lilies especially. We would not have spoken for several shared instants as we inhaled the fragrance of the lilies. I would have enjoyed the staunch companionship of Isaac Gonzales, the watchman who tried to prevent the molasses flood, to have walked through the North End with him and learned about his childhood in Puerto Rico. I would have relished a dainty luncheon—something surrounding a lettuce mound, and definitely dessert—with Fannie Merritt Farmer. And why not a cool viewing of an iceberg with Frederic Tudor, a cup of hot chocolate with Dr. James Baker, and an afternoon of inspiring shopping with Edward A. Filene? These personages have become vivid to me, as I hope they will to readers.

I *have* had lunch with Gayatri Patnaik, senior editor at Beacon Press, who had the idea for this book and, most wonderfully, called upon me to write it. She eats as she edits—with dispatch, elegance, and verve. I am grateful for our association, and for my meetings with P. J. Tierney, possessed of wry humor and joie de vivre, as well as finesse in book production. Tracy Ahlquist, able and adroit editorial assistant, calmly converted scores of motley detached attachments into a virtual book, and did so cheerfully. Lisa Sacks, managing editor, sent letters, notes, and advisories that maintained civil communication and camaraderie during stressful times.

While it is true that I could not have done this book without the contributions of two interns—researchers Melissa Carlson and Elizabeth Steffey—what is more true is that I would not have wanted to do it without them. For this veteran freelance feature writer and essayist, *Boston Firsts* was a personal first. Never before have I labored with assistance, not to mention editorial companionship. I hardly know how I will return to seclusion. Though researching only part-time, Melissa and Liz devoted their thinking toward *Boston Firsts* far more than that, and if the book is appealing and good, I owe it mainly to them. They have been creative, patient, diligent, and more than occasionally brilliant. Several of the ideas for Firsts were theirs. Melissa bit into "Dudley Street Neighborhood Initiative" like a Boston Terrier, and I could not get "YMCA" out of the jaws of Liz. They have kept me honest. Hoy, huzzah, and thank you, gals!

The Boston Public Library has been our stable, stage, and shrine. It is where, in fact, I met P. J. Tierney, production director at Beacon Press, and began a relationship with the press. It is where I visited and lingered over the months, and where Melissa and Liz did most of their research. When the fountain in the courtyard was turned on last summer, we assumed it was mainly for us.

I am indebted to the many writers whose work I have called upon. Every essay subject led to at least a half-dozen major sources, sometimes far more, and there were books that became touchstones. They sit piled beside my desk to reread in a time of detached pleasure. Some concentrated on single subjects, some were literary guidebooks. A smattering include *The Frozen-Water Trade*, by Gavin Weightman, *The Art of Scandal*, by Douglass Shand-Tucci, *Sarah's Long Walk*, by Stephen Kendrick and Paul Kendrick, and the exemplary guides *Boston Sites and Insights*, by Susan Wilson, and *Notorious and Notable New Englanders*, by Peter F. Stevens.

Countless individuals helped me, Melissa, and Liz, especially the staffs of the Boston Public Library (for fast, fast relief from the headache of ignorance, dial 617-536-5400; ask for Reference) and the Massachusetts Historical Society. It is the rare project for which one draws upon the expertise of a specialist in early banana and coconut cargo (historian Dr. Randolph Bartlett of Cape Cod Community College) and also the modern construction of underwater tunnels (civil engineer Charles "Charlie" Button of the Massachusetts Water Resources Authority). Ria Convery, also of the MWRA, is not only a tunnel-nerd, but creative in conveying explanations.

The role of friends during the months of this work could be the basis of another book, about the interwoven fibers of daily support and presence with the more abstract and invisible process of making a book. The two are a fabric of strong rope and dainty thread, jute and silk. *Santé* (and also *shanti*) to Emily Hiestand, my friend of many years, whose presence in my life has been like that of a spring in the woods. She has received me when I am broken, whole, and in between (merely crushed), providing insight, encouragement, and exquisite choices of words. A lesson of midlife: It may be more important to be understood than to be loved. To be both, and to also love, is all.

My pal of thirty years, John Kronenberger, provided technical support; knowledge of English history; lunches of pad thai, Sichuan eggplant, and all-things-vindaloo; and stimulating sartorial presence, a boon to a writer confined to a fashionless room. Susan Pollack, a scribe and near sister, has managed to turn e-mail into a new form, her notes of counsel, encouragement, and congratulation like canoes of cherished supplies sent across waters, arriving on my island. Thank you to my neighbors for inviting me into their lives, especially Lynne Flatley, who plays the computer the way gifted musicians play the piano. And thank you, dear Jack, for months of lilacs, amaryllis, clivia, primroses, peperomia, poinsettia, chrysanthemum, gerbera, dogwood, deutzia, sunflowers, bayberries, and roses. All writers should be presented with baby plants and flowers to accompany their efforts to grow books.

Thank you, dear Boston, most companionable city, and all pilgrims, progressives, and pioneers.